Yuri Avramenko and Andrzej Kraslawski

Case Based Design

Studies in Computational Intelligence, Volume 87

Further volumes of this series can be found on our homepage: springer.com

Vol. 66. Lakhmi C. Jain, Vasile Palade and Dipti Srinivasan (Eds.)
Advances in Evolutionary Computing for System Design, 2007
ISBN 978-3-540-72376-9

Vol. 67. Vassilis G. Kaburlasos and Gerhard X. Ritter (Eds.)
Computational Intelligence Based on Lattice Theory, 2007
ISBN 978-3-540-72686-9

Vol. 68. Cipriano Galindo, Juan-Antonio Fernández-Madrigal and Javier Gonzalez
A Multi-Hierarchical Symbolic Model of the Environment for Improving Mobile Robot Operation, 2007
ISBN 978-3-540-72688-3

Vol. 69. Falko Dressler and Iacopo Carreras (Eds.)
Advances in Biologically Inspired Information Systems: Models, Methods, and Tools, 2007
ISBN 978-3-540-72692-0

Vol. 70. Javaan Singh Chahl, Lakhmi C. Jain, Akiko Mizutani and Mika Sato-Ilic (Eds.)
Innovations in Intelligent Machines-1, 2007
ISBN 978-3-540-72695-1

Vol. 71. Norio Baba, Lakhmi C. Jain and Hisashi Handa (Eds.)
Advanced Intelligent Paradigms in Computer Games, 2007
ISBN 978-3-540-72704-0

Vol. 72. Raymond S.T. Lee and Vincenzo Loia (Eds.)
Computation Intelligence for Agent-based Systems, 2007
ISBN 978-3-540-73175-7

Vol. 73. Petra Perner (Ed.)
Case-Based Reasoning on Images and Signals, 2008
ISBN 978-3-540-73178-8

Vol. 74. Robert Schaefer
Foundation of Global Genetic Optimization, 2007
ISBN 978-3-540-73191-7

Vol. 75. Crina Grosan, Ajith Abraham and Hisao Ishibuchi (Eds.)
Hybrid Evolutionary Algorithms, 2007
ISBN 978-3-540-73296-9

Vol. 76. Subhas Chandra Mukhopadhyay and Gourab Sen Gupta (Eds.)
Autonomous Robots and Agents, 2007
ISBN 978-3-540-73423-9

Vol. 77. Barbara Hammer and Pascal Hitzler (Eds.)
Perspectives of Neural-Symbolic Integration, 2007
ISBN 978-3-540-73953-1

Vol. 78. Costin Badica and Marcin Paprzycki (Eds.)
Intelligent and Distributed Computing, 2008
ISBN 978-3-540-74929-5

Vol. 79. Xing Cai and T.-C. Jim Yeh (Eds.)
Quantitative Information Fusion for Hydrological Sciences, 2008
ISBN 978-3-540-75383-4

Vol. 80. Joachim Diederich
Rule Extraction from Support Vector Machines, 2008
ISBN 978-3-540-75389-6

Vol. 81. K. Sridharan
Robotic Exploration and Landmark Determination, 2008
ISBN 978-3-540-75393-3

Vol. 82. Ajith Abraham, Crina Grosan and Witold Pedrycz (Eds.)
Engineering Evolutionary Intelligent Systems, 2008
ISBN 978-3-540-75395-7

Vol. 83. Bhanu Prasad and S.R.M. Prasanna (Eds.)
Speech, Audio, Image and Biomedical Signal Processing using Neural Networks, 2008
ISBN 978-3-540-75397-1

Vol. 84. Marek R. Ogiela and Ryszard Tadeusiewicz
Modern Computational Intelligence Methods for the Interpretation of Medical Images, 2008
ISBN 978-3-540-75399-5

Vol. 85. Arpad Kelemen, Ajith Abraham and Yulan Liang (Eds.)
Computational Intelligence in Medical Informatics, 2008
ISBN 978-3-540-75766-5

Vol. 86. Zbigniew Les and Mogdalena Les
Shape Understanding Systems, 2008
ISBN 978-3-540-75768-9

Vol. 87. Yuri Avramenko and Andrzej Kraslawski
Case Based Design, 2008
ISBN 978-3-540-75705-4

Yuri Avramenko
Andrzej Kraslawski

Case Based Design

Applications in Process Engineering

With 61 Figures and 23 Tables

Dr. Yuri Avramenko
Lappeenranta University of Technology
Skinnarilankatu 34
FIN-53850, Lappeenranta
Finland
avramenk@lut.fi

Prof. Andrzej Kraslawski
Lappeenranta University of Technology
Skinnarilankatu 34
FIN-53850, Lappeenranta
Finland
Andrzej.Kraslawski@lut.fi

ISBN 978-3-540-75705-4 e-ISBN 978-3-540-75707-8

Studies in Computational Intelligence ISSN 1860-949X

Library of Congress Control Number: 2007938414

Cover design: Deblik, Berlin, Germany

Printed on acid-free paper

9 8 7 6 5 4 3 2 1

springer.com

Foreword

The case-based reasoning (CBR) and case-based design (CBD) have been around for some time and established themselves as one of the commonly used mechanisms of approximate reasoning in intelligent systems and decision support systems, in particular. In a nutshell, the CBR mechanisms offer a powerful and general environment in which we generalize on a basis of already accumulated experience being represented in the form of a finite and relatively small collection of cases. Those cases constitute the essence of the existing domain knowledge. When encountering a new situation we invoke and eventually modify the already collected decision scenarios (cases) and arrive at the pertinent decision or a certain design alternative. Interestingly, uncertainty or granularity of resulting decision is inherently associated with the nature of the cases being used in the reasoning process and a way in which partial matching takes place between the historical findings (cases) and a current evidence.

The book by Professors Avramenko and Kraslawski is unique in several important ways. First, it is an impressive and in-depth treatment of the essence of the case-based reasoning strategy and case-based design dwelling upon the algorithmic facet of the paradigm. Second, the authors provided an excellent applied research framework by showing how this development can be effectively utilized in real word complicated environment of process engineering – a pursuit that is rarely reported in the literature in such a comprehensive manner as done in this book. In a highly authoritative and systematic manner, the authors guide the reader through the essential features of the CBR machinery. The book is structured into 10 chapters. The authors start with some useful generalities by setting up a stage and discussing the principles of the design process of products and stressing on the rapidly growing importance of decision support systems in design activities. Case-based reasoning forms the essence of the consecutive chapter which offers the reader an insight into the algorithms of the reasoning scheme. Chapter 4 is a useful compendium on the variety of concepts that are at the heart of the CBR activities, such as similarity measures and adaptation algorithms. The authors did an excellent job here by combining the badly needed formalism with highly motivating explanatory notes present behind the genuine diversity of the ideas being used there. Chapter 5 brings us closer to the specific applications as the authors navigate the readers through the main functionality of the software environment of the CBR. Finally, the last part of the book consisting of Chaps. 6–9 deals with real-world applications such as, e.g. synthesis of wastewater treatment sequence or design of distillation systems.

While the practicality of the investigations offered in the book is the remarkable strength of the volume, the formal aspects, notation and derivations are rigorous, yet clear.

The writing is lucid and explains the fundamental ideas in a direct manner. The suite of real-word examples is a genuine asset of the book. Through such examples, be they chemical or biological processes, we are provided with a comprehensive, well-structured and clear guidance to the overall design process of the CBR systems.

All in all, the book is an interesting and valuable addition to the body of knowledge on fundamentals and practice of automated reasoning – an important and vitally essential step towards building intelligent systems.

Witold Pedrycz
President, IFSA
May 15, 2007

Preface

The growing amount of knowledge creates new opportunities as well as challenges. Unfortunately, the challenges often start to be the problems. In chemical and process engineering, the most common problems related to the huge amount of available data, information and knowledge are: difficulties with estimation of their quality, lack of efficient methods enabling the fast access to the relevant information or knowledge and "use once" model of knowledge application. The above-mentioned problems are common for all activities in chemical and process engineering: modelling, simulation, design and control. However, design phase is critical from the point of view of the satisfactory functioning of the process unit or the whole system. The wrong assumptions or errors made at this stage could be corrected only with the great amount of time and money but often it is too late for any essential change. The design is difficult as usually there is a lot of uncertainty involved. The good designers used to deal with the problem using their intuition supported by the past experience. The trouble is that industry and society are more and more innovation hungry. There is a growing demand for designs which are less and less similar to their predecessors.

There are two major approaches to deal with this situation, either to make new experiments, develop new models and on this basis build new designs or to use the existing information and knowledge. The second option is much more economically viable and less time demanding than the first one.

The use of the existing information and knowledge is performed in two ways. First method is aimed at getting new information by searching the exiting knowledge repositories. It is so-called knowledge discovery from literature. This approach usually leads to radical innovations. The second method is based on the assumption that the similar problems have the similar solutions. It is a basis of case-based reasoning. It usually leads to incremental innovations.

The objective of this book is to bridge a gap between the huge amount of available knowledge and its very small subset which is not only generated and stored but also actively used. The book is a sort of guide in a store where knowledge is stocked up and we are invited to look for the pieces which could be useful for us in solving new problems. The authors have penetrated only a very small fragment of this huge warehouse – a room in which some elements of knowledge related to chemical and process engineering have been left.

This book is about knowledge re-use by applying of case-based reasoning to the problems typical in chemical product and process design. It is composed of three parts: description of the product and process design and decision support methods related to it, presentation of case-based design principles, issues related to adaptation of the retrieved solutions and case-based reasoning environment and finally examples of application of case-based reasoning to product and process design. The application part covers the broad spectrum of examples dealing with products formulation, synthesis of the system of processing units and mathematical models re-use.

The authors would like to thank many people for the valuable discussions, comments and advice. We are not able to mention all of them but we are particularly grateful to Dr. Tivodar Farkas and Dr. Christan Botar-Jid. We highly appreciate Professor Janusz Kacprzyk for his encouragement and constant support during the preparation of this book.

We hope that this book will contribute to a broader use of case-based design in engineering practice.

Yuri Avramenko
Andrzej Kraslawski
Lappeenranta, May 2007

Contents

Part I Design Support

Part I

Design Support

1 The Design Process of Product and Process Development

1.1 Design Objectives and Tasks

Design is a central activity in chemical engineering, as well as in other engineering related disciplines such as mechanical, electrical, industrial engineering, and computer science. There are common elements in the engineering situation in all these disciplines, and in the way of approaching the goals of design activity. Engineers start designing when there is a need to improve the functions of existing things or to create an artefact with new functions. The most concise and accurate definition of the engineering design activity has been given by Dym and Levitt (1991), who state:

Design is the systematic, intelligent generation and evaluation of specifications for artefacts whose form and function achieve stated objectives and satisfy specified constraints.

Design activity must start with a goal, constrains within which the goal must be achieved and criteria by which the solution might be recognized. The result of the activity is a detailed description of an artefact provided by a set of specifications. Dym and Little (2004) defined the specifications of artefacts as "precise descriptions of the properties of the object being designed". This description must be sufficient for the realization of the artefact.

Design requirements the characterisation of the perceived needs around the artefact environment. These perceived needs are transferred to the goals. In order to work practically with goals, they need to be characterised into one or more statements. Any characterised statement about a goal is called an objective. A design requirement is an objective that has to be met by the design (Roozenburg and Eekels, 1995).

Based on this description of design activity, the design of chemical processes begins with the desire to produce chemicals that satisfy certain needs. The identification of the function properties of the chemical product that correspond to the needs and their transition to structural properties of the product (product specification) can be regarded as product design in chemical engineering.

Y. Avramenko and A. Kraslawski: *Case-Based Design*, Studies in Computational Intelligence (SCI) **87**, 3–24 (2008)
www.springerlink.com

Process design establishes the sequence of chemical and physical operations, operating conditions, duties, and specification of all process equipment; the general arrangement of equipment needed to ensure proper functionality. The process design is summarized by a process flowsheet, materials and energy balances, and a set of individual equipment specifications (Walas, 1988).

Designing an artefact can be considered a transition from concept and ideas to concrete descriptions. The design specifications, which include the constraints of a design problem, may initially not be precise or complete. In addition, alternative design solutions are not available in advance and must be developed by a specific research process. The solution of the design task is usually evaluated by satisfactory criteria rather than finding of optimal solution. These characteristics of design activity are true of chemical process design as well.

Design tasks can be classified into three categories: routine, innovative, and creative (Brown and Chandrasekaran, 1985; Gero, 1990). In routine design all variables and their application ranges, as well as the knowledge to compute their values, are directly derivable from existing designs. Routine design problems are typically represented by a well defined set of components and a set of constraints that the final design must satisfy. The task of this design activity is usually to find the appropriate alternatives for each subpart that satisfies the given constraints. In contrast to routine design, innovative design tasks are usually described by incomplete knowledge, and the applicable range for variable values may change. The result of innovative design is a novel design with a familiar structure but unfamiliar set of values of the defined variables and their combinations. The design might be an original combination of existing components. As part of innovative design, redesign of artefacts takes place when the artefact fails to satisfy one or more critical new requirements, or the environment for which the artefact had been designed changes (Braha and Maimon, 1998). Creative design can be defined as non-routine design that introduces new variables and, as a result, extends or moves the space of potential designs. In creative design the set of possible solutions is unknown.

Modern process designs are rarely routine; rather, they involve innovative approaches to integrated processes that are more profitable, as well as easily controlled and environmentally safe.

A classification of design tasks is needed to facilitate the organization of the knowledge, representation, and reasoning schemes that would be useful in supporting different kinds of design.

1.2 Design Models

Design has been discussed in contexts such as general design methodologies (Cross, 1984; Dasgupta, 1989; Braha and Maimon, 1998), design artefact representation (Rinderle, 1987; Surma and Braunschweig, 1996) and computational models for the design process (Rivard and Fenves, 2000). Many design researchers declare that the design process is stepwise, iterative and evolutionary.

1.2.1 Generic Design Models

There have been many attempts to develop models of the design process. Models of the design process are often drawn in flow-diagram form, where the design process proceeds from one stage to another with feedback showing the iterative returns to earlier stages.

One of the simplest of the design models consists of four steps (Fig. 1.1). In the first stage, exploration, the objectives are clarified, the design requirements are established and the constraints are identified. In the second stage, the generation of the design proposal takes place. The design proposal is subject to evaluation against objectives, constraints and criteria in the evaluation stage. Refining and possible optimization of the design is done in this stage as well. Documenting the design proposal and its communication to the manufacturers are performed in the final stage, communication (Cross, 2000).

French (1985) has developed a more detailed model of the design process (Fig. 1.2). In the flowchart, the circles represent the data context of the design stages, and the rectangles indicate the design activity. The process begins with an initial statement of a need and the first design activity is analysis of the problem. The output is a statement of problems composed of clarified objectives, constraints and evaluation criteria. The activity that follows is conceptual design, where the solutions which can be used to solve the stated problems are generated in the form of concepts. The output of the conceptual design stage is a set of possible concepts, or schemes, for the design.

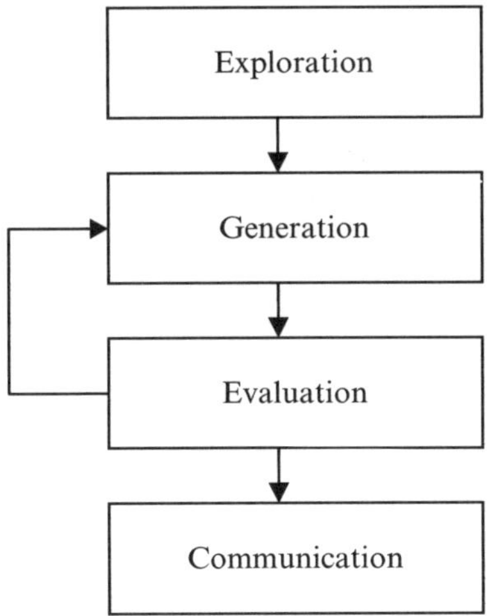

Fig. 1.1. Four stages model of design (after Cross, 2000)

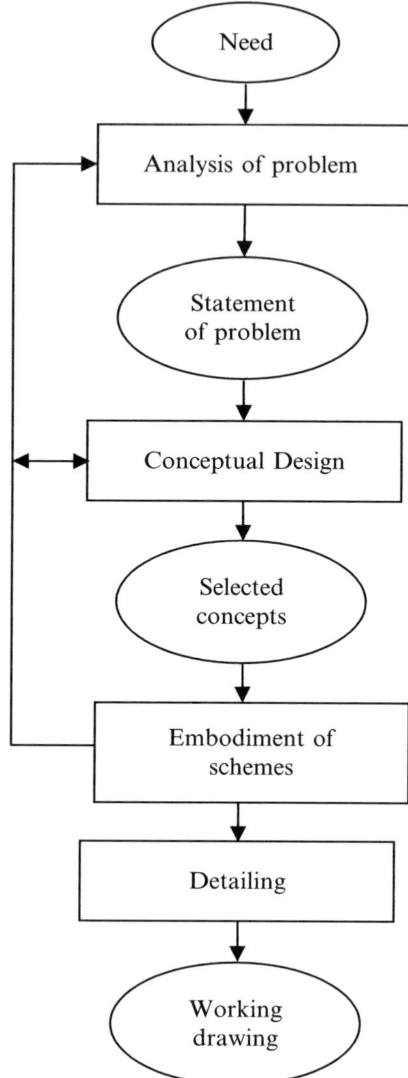

Fig. 1.2. French's design model (after French, 1985)

According to French (1992), a scheme is "an outline solution to a design problem, carried to a point where the means of performing each major function has been fixed, as have the spatial and structural relationship of the principal components". The next stage of the design process is the embodiment of the concepts. The conceptual proposals are detailed and then evaluated; a final choice between different schemes is made. In this stage the selection and sizing of the subsystems is done, based on lower-level concerns that include the performance specifications and the operating requirements. The last step is detailing, in which a very large number of small but essential points remain to be decided.

A more complex design model has been proposed by Pahl and Beitz (1984) (Fig. 1.3). The model includes four stages: clarification of the tasks, conceptual

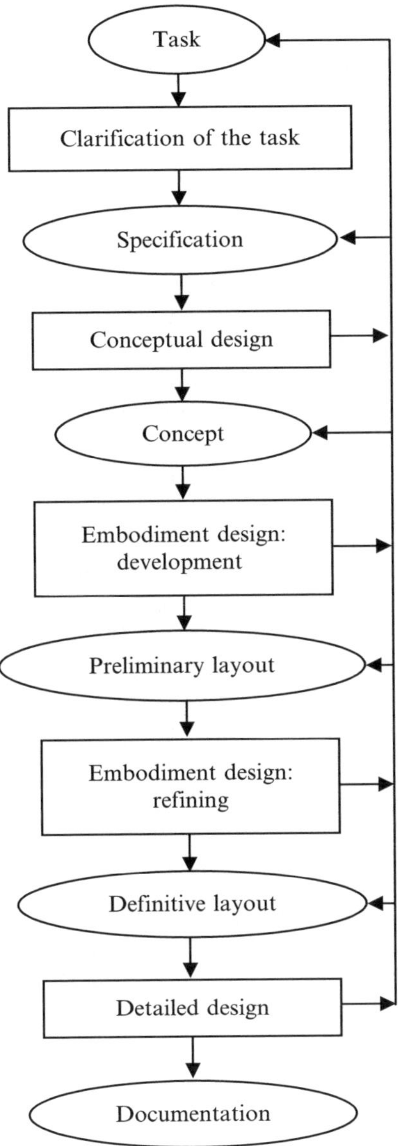

Fig. 1.3. The model of design process (after Pahl and Beitz, 1984)

design, embodiment design and detailed design. The first stage is collection of information about the requirements and design objectives and the establishing of design specifications and constraints. Conceptual design establishes function structures, searches for suitable solution principles and combines them into concept variants. The principle functions in this stage might be decomposed into sub-functions that can be performed by individual components or subsystems. The embodiment design stage of the design process has two sub-stages; the first produces a preliminary layout obtained by refining the conceptual designs, evaluating them against technical and economic criteria and selecting the best one; the second has as its output a definitive layout after accomplishing the preliminary layout and testing for errors and effectiveness. The detailed design finalizes the layout, checks technical and economic feasibility, produces manufacturing specifications and gives as outputs the final documents.

The stages are often carried out iteratively, returning to preceding ones, providing the feedback and ability for improvement.

The stages before generation of the preliminary layout can be considered the procedure of optimization of the principle, while the three last stages deal with optimization of the layout and forms.

The French's classic model of the design process has been extended by Dym and Little (2004) to a five-stage model (Fig. 1.4). The model defines what is done in each stage by incorporating the design tasks of stages.

Each stage requires an input, has design tasks that must be performed, and produces an output. The stage tasks are supplemented with sources of information. During problem definition the design objectives are clarified and information needed to develop an engineering statement of functional needs is collected from literature, experts, and regulations. The stage has as its output refined objectives, constraints, requirements and functions. The next stage, conceptual design, generates concepts or schemes of designs alternatives. In the preliminary design stage, identification of the principle attribute of the design concepts is carried out. The sources of information include heuristics, simple models and known physical relationship. The concepts are analyzed and evaluated. The selected design proposal proceeds to the following stage. Detailed design refines the design proposal, details it and proposes the manufactory specifications.

Handbooks, local laws, and suppliers' component specifications serve as the source of information for the detailed design. Finally, during the design communications stage the manufactory specifications and their justifications are produced. Iteration and feedback are integrated in all stages of the design process (not shown in Fig. 1.4).

1.2.2 Chemical Process Design Model

Design of a chemical process follows almost the same steps. Chemical engineers designing the new chemical product face two major tasks: determination of

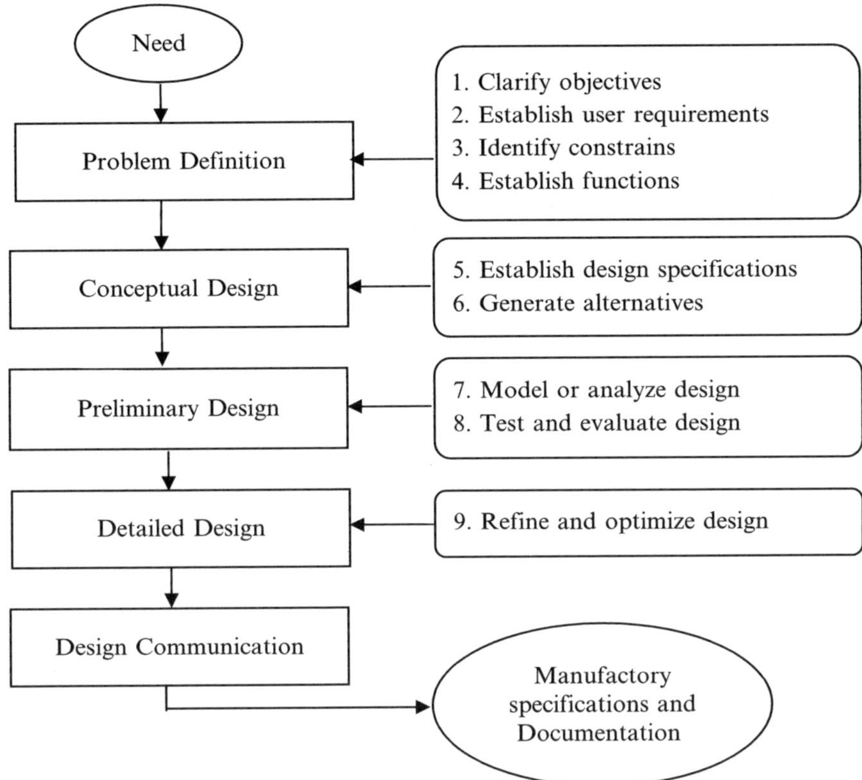

Fig. 1.4. The five-stage design model (after Dym and Little, 2004)

the composition of the chemical mixture or structure of the new chemical compound to provide the desired properties, and creation of process flowsheets with operating conditions to produce the desired products with a high degree of yield and selectivity.

Seider et al. (1999) suggested a general model of design for chemical processes. It begins with a potential opportunity, which can be understood as a claim to new chemicals with desired properties, the availability of an inexpensive source of raw materials, or confidence that a new route of production of an existing chemical may be profitable. At the first stage, identification of needs and generation of ideas, the primitive problem is created and assessed. When necessary, a search for chemicals or chemical mixtures that have the desired properties and performance is carried out. The identification of the product specifications is done in this stage. Then, the process creation phase occurs. It is composed of a preliminary process synthesis and detailed process synthesis. During the preliminary process synthesis, reactions, separations, temperature and pressure changes, operations, task integration, and operating mode (continuous, batch, or semicontinuous) are considered. This stage can be regarded as conceptual process synthesis. Detailed process synthesis comprises such tasks as synthesis of chemical reactor networks, separation train synthesis, second law analysis, synthesis of heat exchanger networks, synthesis of mass exchanger networks. The result of the detailed process synthesis stage is promising flowsheets that deserve to be developed further. The detailed design of these alternatives involves equipment sizing of units, cost estimation, profitability analysis and optimization. Identification of the required additional equipment is undertaken. Another activity considered in this stage is analysis of the reliability and safety of the proposed process.

When the detailed process design stage has been complete, the economic feasibility of the process is checked to meet profitability requirements. Then an assessment of the flowsheet controllability is initiated that begins the qualitative synthesis of the control structure for the entire flow diagram. The final stage may be considered as the plant design. The complete equipment drawings and layouts, piping diagrams, instrumental diagrams, and construction are the subject of the final design stage. A flowchart summarizing the chemical design process is given in Fig. 1.5.

1.2.3 Product Design Models

Many chemical products are manufactured in small quantities and the design of a product focuses on identification of the chemicals or mixture of chemicals that have the desired properties, such as stickiness, porosity, and permeability, to satisfy specific industrial needs (Seider et al., 2004).

In the previous chemical process design model, the identification of the product specifications is considered as an initial part of process design. However, many researchers differentiate between the activities performed in product and process designs.

Product design can be defined as the idea generation, concept development, testing and manufacturing or implementation of a physical object or service. A generic product development process has been described by Ulrich and Eppinger (2000). The product design model comprises seven steps (Fig. 1.6):

- Identifying needs, which has the goal of clarify the needs. The output is a problem statement.
- Establishing product specifications, that is a precise description of product functionality in technical terms. The specifications might be refined to be consistent with the constraints.
- Concept generation, where the space of ideas for the product is explored and product concepts that satisfy the needs are produced. Concept generation includes the search for, creative problem solving, and the systematic exploration of various possible options. The result of the activity is a set of product concepts.

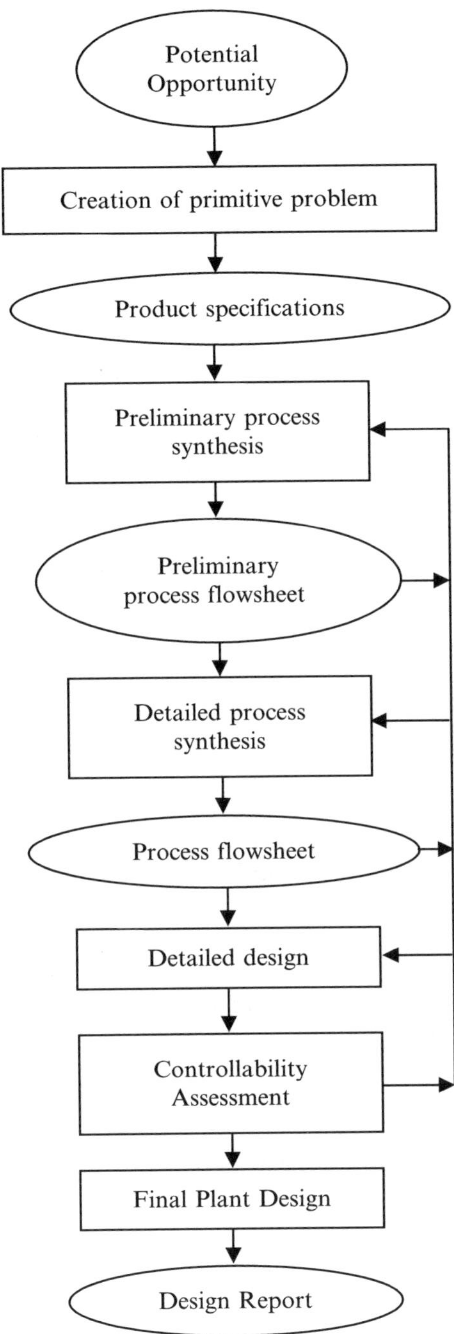

Fig. 1.5. The chemical process design model (modified Seider et al., 2004)

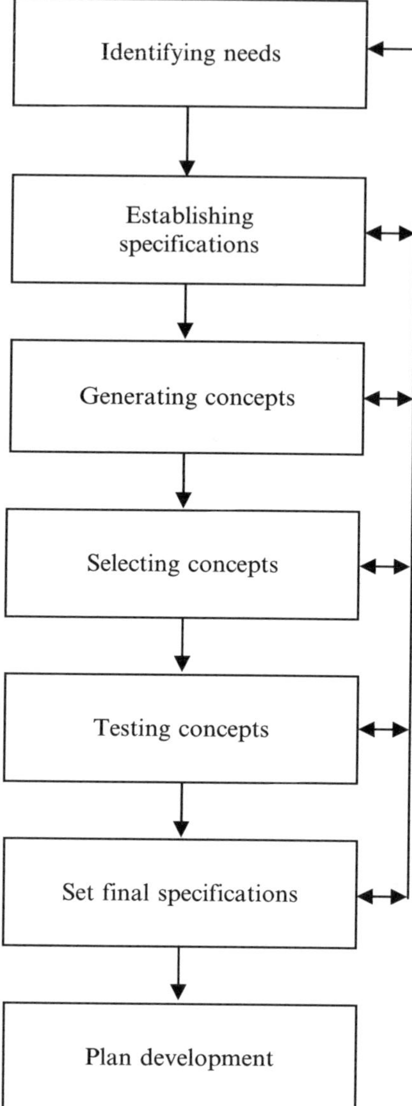

Fig. 1.6. The product design model (after Ulrich and Eppinger, 2000)

- Concept selection, the activity in which various product concepts are analyzed and sequentially eliminated to choose the most promising concept. The step usually requires several iterations and may initiate additional concept generation and refinement.
- Concept testing, in which one or more selected concepts are tested to verify that the needs have been achieved.

- Setting final specifications, where a technical model of the product is developed to get precise values for product properties.
- Plan development, this step comprises development of a strategy to minimize development time, identifying the resources required to complete the project, and creation of the production plan.

It is not necessary that the design process proceeds in a sequential way, where each step is completed before beginning the next. The activities of steps might overlap, and iterations occur. This behaviour is shown in the flowchart of deign process by means of two-way arrows.

Cross (2000) proposed another model for product design, which consists of seven stages and covers all aspects of the design process from problem clarification to detail design, applying a systematic approach.

Since at the beginning of the design process of a product a complete and clear statement of design objectives is rarely given, Cross positioned the clarification of the design objective as an important fist step of the product design. The identification of sub-objectives and relationships between them are performed at this stage. The next step focuses on what has to be achieved to be a new design. The essential functions which the product to be designed must satisfy no matter what physical components might be used are determined at this stage. The set of requirements comprising the performance specifications of the product is the subject of next step of the design process. Specification limits are also set at this stage. In the following stage, the set of targets to be achieved by the engineering characteristics of a product, such that they satisfy the requirements, with their relative importance, are determined. The generation of possible product concepts is the next step. The aim of the stage is to create a complete range of alternative design solutions for the product and hence to widen the search for potential new solutions (Cross, 2000). Then, the utility values of the alternative design proposals are compared on the basis of performance against objectives. The last step is adjusting the product details to increase its performance. A flowchart of the design process described is presented in Fig. 1.7.

Wibowo and Ng (2001) presented a model for product design, and applied to such chemical products as creams and pastes (Fig. 1.8).

The first step in their design process involves the identification of product quality factors that satisfy the needs. The quality factors are divided into four groups: functional (protect, clean, decorate the body, deliver the ingredient, etc.), rheological (spread easily on the skin, coats uniformly, etc.), physical (remain stable for an extended period, melt at a specific temperature, etc.), and sensorial (smooth feeling, opaque, do not cause irritation, etc.).

Given these factors, the second step is product formulation, which involves the selection of ingredients, the emulsion type and determination of product microstructure. Then, the design of the production process and product evaluation steps follow.

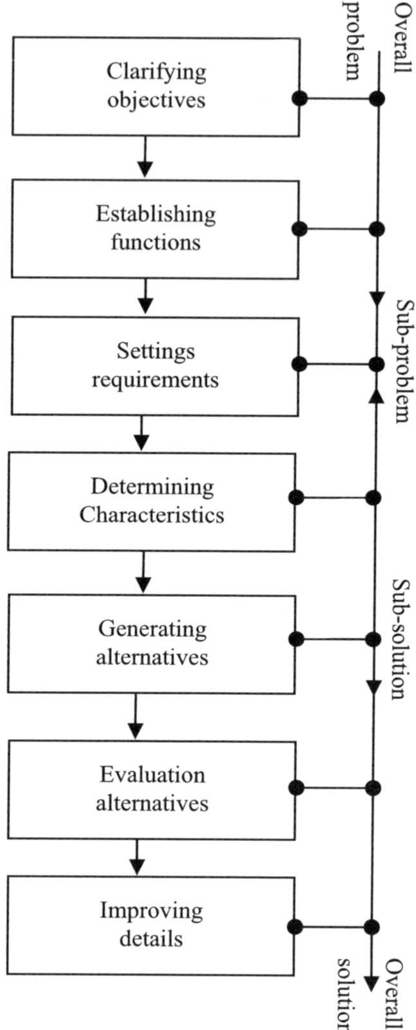

Fig. 1.7. Cross's product design model (modified Cross, 2000)

Cussler and Moggridge (2001) introduced a generalized model for chemical product design, which is most cited nowadays. They divided the design procedure into four steps: defining needs, generating ideas, selecting the most promising ideas, and manufacturing the product (Fig. 1.9).

Each step is decomposed according to the sub-tasks performed. Identifying needs involves interpretation of needs, conversion of the needs to specifications, and revising the product specifications. Redundant and irrelevant needs are refined to be more accurate; they are ranked to essential, desirable

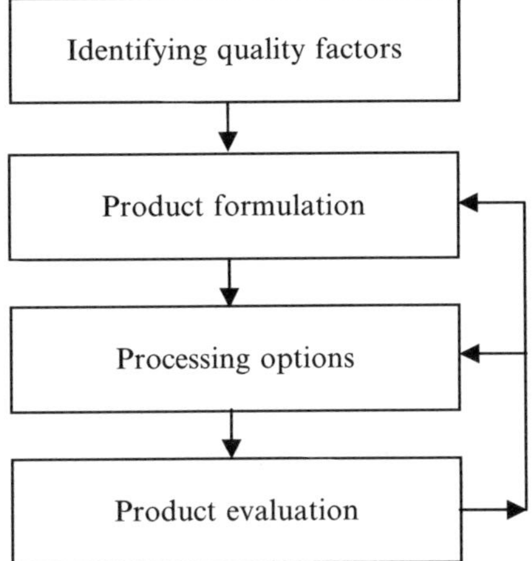

Fig. 1.8. Cosmetic product design model (after Wibowo and Ng, 2001)

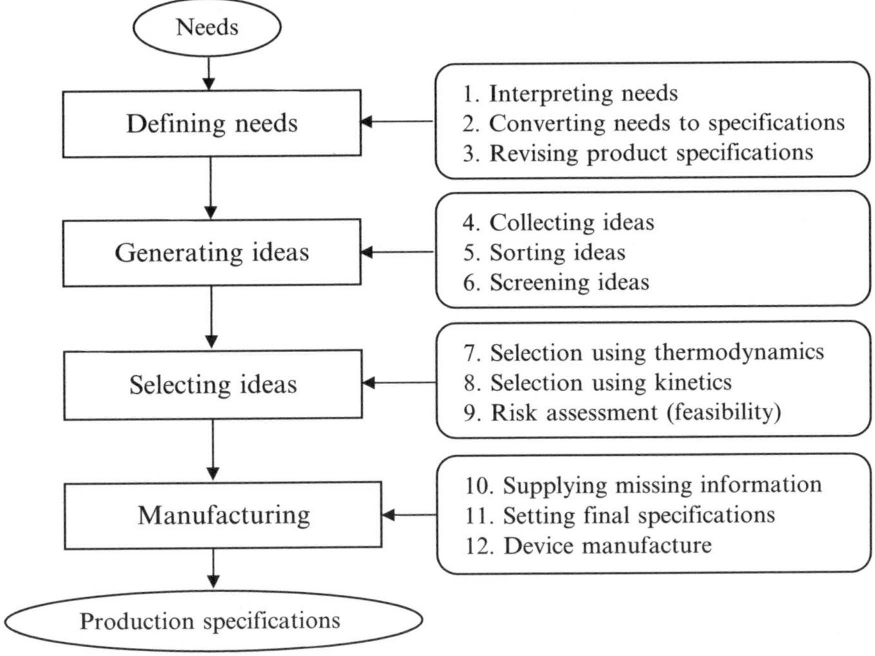

Fig. 1.9. Chemical product design model (modified Cussler and Moggridge, 2001)

and useful groups. Given an array of needs, an effort is made to convert qualitative needs into particular specifications for the product. The preliminary specifications require revision and the revised specifications must be analysed critically to see whether they make sense.

The generation of ideas comprises collecting the ideas from various sources and sorting the ideas by different methods where only relevant ideas are kept and screening the ideas using a concept-screening procedure to judge the advantages and disadvantages of the ideas and finally reducing the number of concepts for more quantitative consideration.

Having generated ideas, an attempt is made to select the most promising among them using the principles of thermodynamics and chemical kinetics, as well as feasibility analysis.

In the last step of chemical product design it is essential to provide a complete and rigorous description of the selected concepts. Discovering concept details might require further research and experimentation. One form of information, commonly required for the design of chemicals, is a synthetic pathway for the active molecules. Specifying the product structure involves four items: chemical composition, physical geometry, chemical reactions, and product thermodynamics. This stage is the point where the chemical process design normally begins (Cussler and Moggridge, 2001).

It can be stated that finding product properties (physical and structural) with a certain functionality is the key activity of chemical product design, while creation of a production process description is the problem of chemical process design.

1.2.4 Product Design Models Based on Abstraction Levels

Among other models of the design process, there are notable ones which consider the design process as an iterative process of concretion performed on different levels of abstraction. This approach has been developed for mechanical design problems.

The researchers belonging to the German engineering design school describe mechanical design categorized on modelling layers representing certain levels of abstraction (Grabowski et al., 1995). These different levels of abstraction are classified into:

- *The requirements* modelling layer, containing the preconditions of the design, the definition of the product requirements and the description of the product's task structure;
- *The functional* modelling layer, serving to represent the functions and the functional interrelationships of the design artefact to be developed;
- *The conceptual* modelling layer, containing all information to describe the solution concept of a design, such as physical solution principles, effective spaces, faces and lines and the grouping of the functional structure into a conceptual structure;
- *The shape* modelling layer, completing the above product modelling layers by giving the geometrical definitions to fully defined three-dimensional parts with assigned material properties and their combination into a part group structure.

Dixon and his colleges (Dixon et al., 1988) suggested a slightly different classification of abstraction levels in regarding mechanical design problems. The six levels are:

- *Perceived needs*, the initial conditions, constraints, requirements and goal of the design; it provides the motivation for designing the product.
- *Functional*, a more detailed statement of the needs, without reference to physical principles and form. A function performance of a product is translated into a detailed, quantitative, operational statement of functional requirements.
- *Phenomenological*, the level of working principles, physical phenomena which underlie the designing of the artefact.
- *Embodiment*, at which a generalized form or shape based on the physical phenomena being developed to achieve the functions is represented.
- *Attributive*, where the specific attribute types, the product type are detailed.
- *Parametric*, at which the specific values of the product attributes are given.

The design process is characterized as a series of stages, each of which transforms the initial state, a higher abstraction level, to a final state, the lower abstraction level. For example, the attributive stage defines as a final state a class of product being designed and the types of its parameters, whereas the parametric stage gives a list of values of the product parameters.

1.2.5 Summary of Design Process Models

After review of the different models of the design process proposed by design researchers it can be stated that design is a sequential process with iteration loops between or within stages. The design process evolves from idea to realization. During the design progress, a new idea or new technology may become available that requires modification of the initial design proposal. Often, the designer must iteratively break down the set of requirements into dimensions, constraints, and features and then test the resulting design to see if the remaining requirements were satisfied.

One assertion can be extracted from the design models: the specification of intermediate outputs of the design process increases after each step. Starting with a very abstract description of needs, the design process evolves to a final, very detailed and specific proposal.

1.3 Model of the Design Process for the Development of a Chemical Product

The general intention of any design activity in chemical engineering leads to the statement: how to produce a chemical product with the desired behaviour and functionality. Even if the intention of the design is to modify the existing production process without changing the final product type, or to develop a newly composed product with an unchanged production line, the statement above remains correct.

It would be desirable to build a general concept of design activity in chemical engineering towards supporting the decision making process with computer tools. In order to be useful the concept must contain a general representation of the design activity and provide a clear description of the design process at every stage. The description of atomic design stages lies within the traditional analysis, synthesis and evaluation design paradigm, while the general outline of the design process represents the evolution of reducing abstraction in artefact description.

1.3.1 Representation of a Design Activity

According to the definition of the design given in the beginning of the chapter, the design activity is a goal-directed derivation process which starts with an abstract description of an artefact and ends with its more detailed description and follows such actions as

- *Analysis* of statement
- *Generation* of options
- *Evaluation* of solutions

Analysis means understanding the initial statement, perceiving goals, and defining the strategy of concretisation of the initial statement. By generating options, all possible ways of concretisation of the statement are considered. Options are not solutions; they may contain a solution, but it might not be realistic. The evaluation action refines the options to retrieve the solutions and involves assessment to select the most proper option which represents a way of achieving the goals and creating the final statement.

This is elementary design activity. However, real design processes may require several actions for concretisation of the artefact description. A complex design process is a combination of elementary design activities. The general direction of the design is from the more abstract to the less abstract levels. If the evaluation procedure fails all solutions of the stage, then the current position of the design process is lifted to the previous level of abstraction. Thus, there are two directions of the design process: consecutive, following all elementary actions of the design activity to a lower level of abstraction, and stepwise, to a higher abstraction level (Fig. 1.10).

1.3.2 Overall Process of the Design of a Chemical Product

The design process starts with the need to produce a chemical product with a given behaviour. The final statement of the design must be the detailed description of the production process of the desired product.

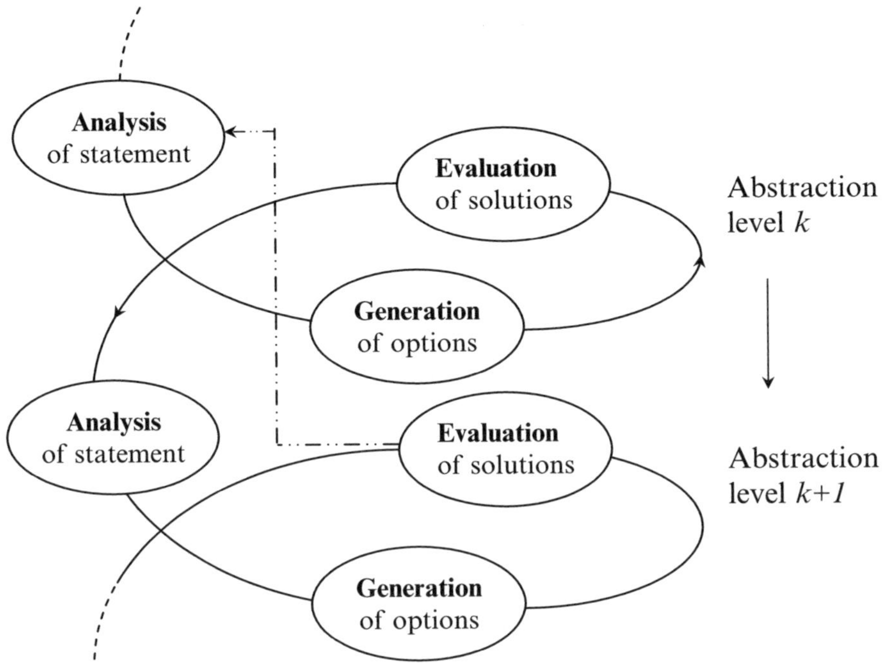

Fig. 1.10. The directions of the design process through abstraction levels

The starting point for a chemical product design is very often an ill-defined problem statement. It is quite rare for a designer to be given a complete and clear statement of design objectives. There is a need for clarification of the design objectives and formulation of the problem statement. This must be the first step that precedes any further design activity. But the problem definition step is not an elementary design activity, as it was described in the previous section. It is regarded as pre-design stage (e.g. Gani, 2004).

The process of design must start with clear formulation of design objectives and the defined design problem. The pre-design phase must include the identification of the functionality of the desired product which reflects its behaviour.

The functionality of the product being designed is represented by quality factors. These quality factors are translated to the physical–chemical properties (such as viscosity, density, refractive index, solid fat index, etc.), which the product must have. These properties, called quality variables (Bernard and Saraiva, 2005), are related to the functionality of the product. Within the developed concept of the design process, the identification of the physical–chemical properties of the future product is based on a functionality representation called the Properties Design.

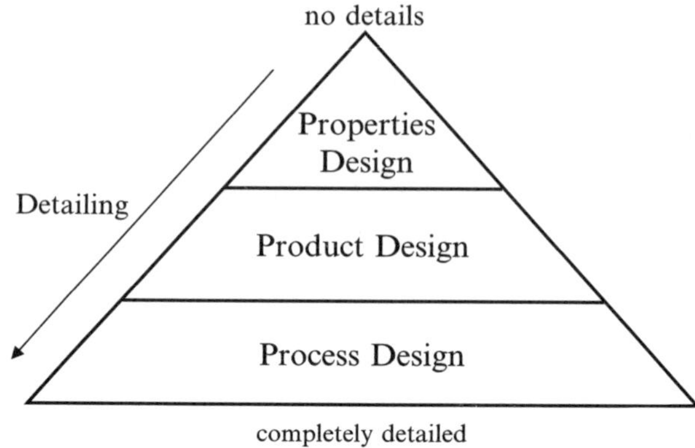

Fig. 1.11. The stages of the design process for the production of a chemical product

The next design activity is determination of the structural properties of the product. Either the molecular structure of the chemicals or the composition of the mixture/blend and colloidal system is defined. This stage is the Product Design.

After the chemical formulation has been identified, it is often necessary to design a manufacturing process, which is the consideration of the Process Design.

Thus, the overall design process is represented as three PROs – Properties design, Product design and Process Design (Fig. 1.11).

The detailing of the design increases from Properties design to Process design.

Product design is an elementary design activity. It is also composed of analysis of physical properties and constraints, generation of product structures and then evaluation and selection the most promising ones (that conform in general to a known chemical product design model, e.g. Fig. 1.9). It can be a molecular structure design where numerous permutations of atoms and molecular groups are performed to achieve the desired properties of a compound. The search for molecular structure is often iterative, involving heuristics, experimentation, and the need to evaluate numerous alternatives (Seider et al., 2004). Product design can also involve the design of mixtures or blends, where the proper structure and composition of molecular systems (consisting of one, two or three phases) are identified, and solvent selection (when the design of a new solvent is a problem for molecular structure design).

Process design is a more complex design activity and involves several levels of abstraction of the design statement. The process design can be described as a sequence of elementary design activities leading to intermediate results.

1.3.3 Abstraction-Level Based Model of Chemical Process Design

The design process begins with *objectives* (desired products) and ends with *realization of objectives* (process flowsheet with equipment specifications).

The first step is creation of a completely abstract description of the process and no unit operations and equipment specifications are considered. It then passes through intermediate steps, and finally arrives at a final design where all equipment and operation parameters are completely determined. The reasoning process requires an iterative process in which the level of abstraction in incrementally refined to establish a solution. Each solution reduces the design space and serves as a starting point for a local search at a more concrete level.

The design engineer might begin the design of a chemical process by sketching out a simple diagram in which only the feed and product streams are identified. Then the engineer might break down the process into its basic functional elements such as reaction, separation, heat exchange. The engineer would also consider proper processes as the realization of functional blocks. After complete material and energy balances performed and preliminary equipment parameters determined the process diagram becomes more complicated. Finally, the mechanical and instrumental details of the process are considered and equipment specifications are defined. In addition, the economical and environmental parameters could be estimated.

The overall design process is divided into the following design activities (Fig. 1.12):

A. Abstract Design

This design activity is independent of equipment and unit operations. It defines the principle way of production of the desired product(s). The chemical reactions, if any (stoichiometry and kinetics, catalysis) are subject to consideration. Unwanted side reactions must be considered as well. Only process streams are identified, utility streams are not considered. Individual components flows can be manipulated to get the good conditions for the process. The result is examined to evaluate its feasibility.

Result: principles of process, how to get the product from raw materials – Input–Output diagram including reactions pathways, conditions, selected catalysts types.

B. Basic Design

As the first action in this abstract-level, the states of raw materials, intermediates and final products are determined. The state is defined by mass, composition (mole or mass fractions), phase (solid, liquid, or gas), form (if solid), temperature and pressure.

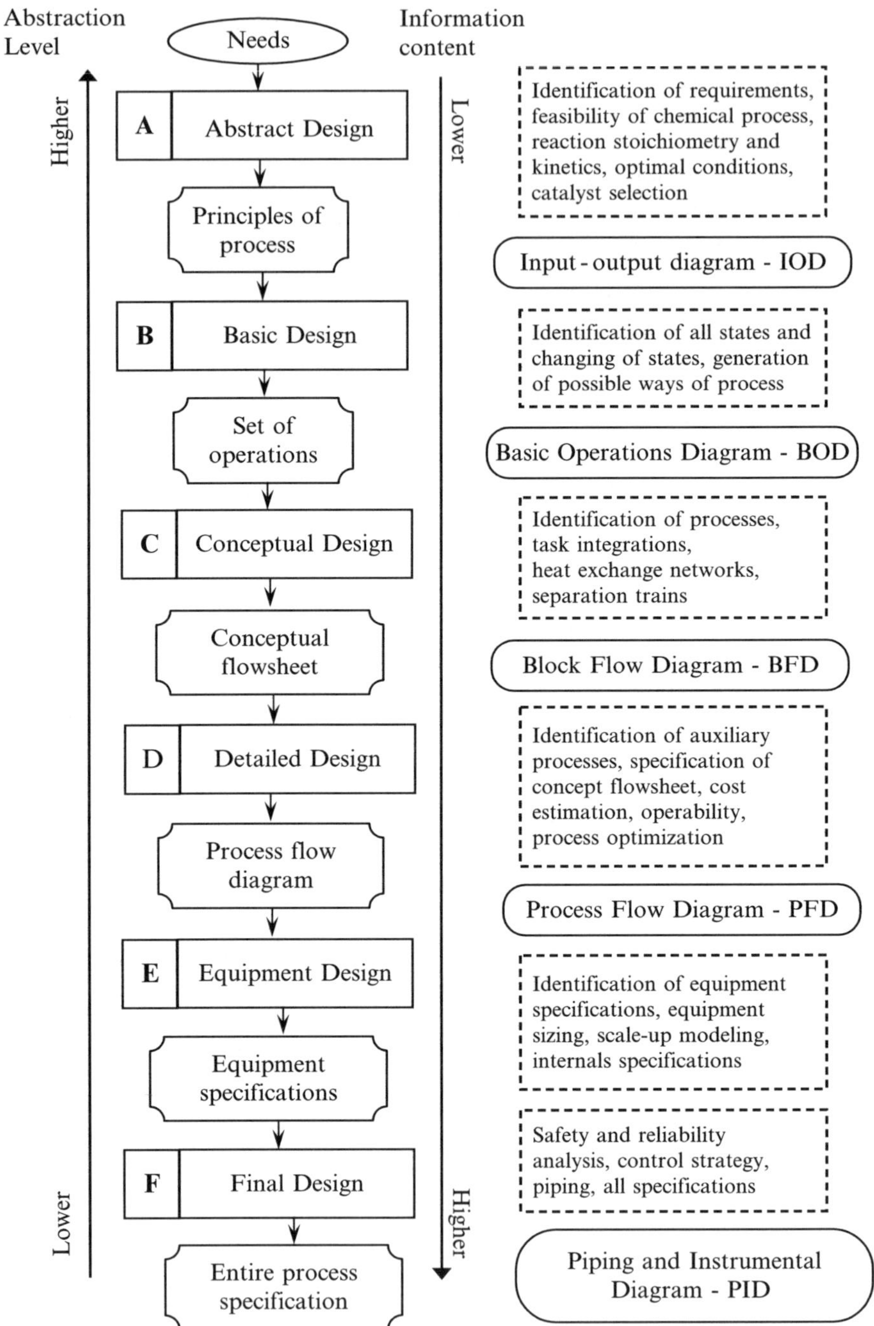

Fig. 1.12. Chemical process design model based on abstraction-levels

Analyses the possible ways of processing the set of basic physical–chemical operations are created. The basic operations are the building blocks of nearly all chemical and physical processes. They are:

(1) Chemical reaction
(2) Separation of mixtures
(3) Phase separation
(4) Temperature change
(5) Pressure change
(6) Phase change
(7) Mixing and splitting of streams
(8) Form changing (if solids)

An idea of which types of operations are best for the process can be obtained. Where feasible, combining some basic operations is considered.

The basic operations by are united into areas realized certain functions in a process being designed. These function blocks can be, for example, reactor feed preparation, reactor, separator feed preparation, separator, recycle (Turton et al., 1998).

Result: the alternatives of sequences of basic operations – basic operations diagrams.

C. Conceptual Design

Given the states of components and process streams, this design activity considers the selection of process operations. This stage is still not concerned with specific equipment and operation parameters. Only types of processes, connections, and input/output streams are represented in the block diagram. A decision on utilizing batch or continuous process types is made. Mixing process, heat exchange networks and separation trains with possible task integration are subject to be designed.

Result: Flowsheet involving selected process operations (process types).

D. Detailed Design

This design activity provides a more detailed view of the process. It determines a map of equipment, their specifications and operation parameters. Utility streams are specified. It also involves cost estimations, operability, and process optimization.

Result: process flow diagram representing all equipment, connections, main control loops, providing streams information, and operating conditions (pressure, temperature, flow rates).

E. Equipment Design

This step involves equipment selection, equipment sizing, and design of novel equipment.

Result: Equipment specifications.

F. Final Design

At this stage the entire technological scheme is generated, all process-specific information is supplied, and all economic and ecological calculations are performed.

Result: Final documentation.

In many cases the A and B steps are merged to one step: ABstract design, where the principles of the process are considered together with possible ways of solving the stated problem. When a new type of equipment or novel enhancement of conventional equipment is not required, the E step can be integrated to the D step to produce one DEtailed design phase. In some cases, even detailed construction of a new process is not required. For instance, such chemical products as cream and pastes need to precisely identify the structure of product, but the process of production of the defined structure is nearly known. When the structure has been identified the process concepts (way of productions) are generated, and after evaluation the selected one is specified. Thus, process design is compacted to only two stages: ABstraCtive and DEFinitive designs.

The result of the above activity is a process design, described in the form of a process flow diagram showing the individual process steps that can generate desired products from available raw materials under appropriate operation conditions, and which are interconnected in a certain way.

2

Decision Support in Design

Design is a problem solving activity. Decision making during the design activity deals with highly complex situations. Decision-making methods can be applied as techniques that are able to assist the designer in the design process.

2.1 Decision Making Process

Making a decision implies that there are alternative choices to be considered, and in such a case the goal is not only to identify as many of these alternatives as possible but to choose the one that best fits with specified objectives (Harris, 1998).

For most familiar everyday problems, decisions based on intuition can produce acceptable results because they involve few objectives and only one or two decision-makers. In the engineering environment, problems are more complex. Most decisions involve multiple objectives, several decision-makers, and are subject to external review. The specific methods for decision support are the key aspect in design practice.

A general decision making process can be divided into the following steps:

1. Problem definition
2. Requirements identification
3. Goal establishment
4. Evaluation criteria development

The process may return to a previous step from any point in the process when new information is discovered. Thus, this repeats most of model of design process, which is virtually the decision-making process.

2.1.1 Definition of the Problem

Problem definition is the crucial first step in making a good decision. This process must, as a minimum, identify root causes, limiting assumptions, system and organizational boundaries and interfaces, and any stakeholder issues.

Y. Avramenko and A. Kraslawski: *Case-Based Design*, Studies in Computational Intelligence (SCI) **87**, 25–48 (2008)
www.springerlink.com

The goal is to express the issue in a clear *problem statement* that describes both the initial conditions and the desired conditions. The problem statement must however be a concise and unambiguous material. It is essential that the decision-maker ensures what problem is going to be solved before proceeding to the next steps.

Result: Problem statement – functions, initial conditions, desired state etc.

2.1.2 Identification of Requirements

Requirements are conditions that any acceptable solution to the problem must meet. Requirements represent what the solution to the problem must do. For example, a requirement might be that a process must produce at least ten units per day. Any alternatives that produced only nine units per day would be discarded. Requirements that do not discriminate between alternatives need not be used at this time.

In mathematical form, these requirements are the constraints describing the set of the feasible (admissible) solutions of the decision problem. It is very important that even if subjective or judgmental evaluations may occur in the following steps, the requirements must be stated in exact quantitative form, i.e. for any possible solution it has to be decided unambiguously whether it meets the requirements or not.

Result: List of absolute requirements.

2.1.3 Establishment of Goals

Goals are broad statements of intent and desirable programmatic values. Examples might be: reduce worker radiological exposure, lower costs, lower public risk, etc. Goals go beyond the minimum essential requirements to wants and desires. Goals should be stated positively (i.e. what something should do, not what it should not do). In mathematical form, the goals are objectives contrary to the requirements that are constraints. Because goals are useful in identifying superior alternatives (i.e. define in more detail the desired state of the problem), they are developed prior to alternative identification.

The goals may be conflicting but this is a natural concomitant of practical decision situations. During goal definition, it is not necessary to eliminate conflict among goals nor to define the relative importance of the goals. The process of establishing goals may suggest new or revised requirements or requirements that should be converted to goals. In any case, understanding the requirements and goals is important to defining alternatives.

Result: List of clearly formulated goals.

2.1.4 Generation of Alternatives

Alternatives offer different approaches for changing the initial condition into the desired condition. Be it an existing one or only constructed in mind, any alternative must meet the requirements. The decision team evaluates the requirements and goals and suggests alternatives that will meet the requirements and satisfy as many goals as possible. If the number of the possible alternatives is finite then it is possible to check one by one for meeting the requirements. The alternatives vary in their ability to meet the requirements and goals. If an alternative does not meet the requirements, three actions are possible:

(1) The alternative is discarded.
(2) The requirement is changed or eliminated.
(3) The requirement is restated as a goal.

The infeasible alternatives must be deleted from the further consideration, and the explicit list of the alternatives is generated. If the number of the possible alternatives is infinite, the set of alternatives is considered as the set of the solutions fulfilling the constraints in the mathematical form of the requirements.

The description of each alternative must clearly show how it solves the defined problem and how it differs from the other alternatives. A description and a diagram of the specific functions performed to solve the problem will prove useful.

Result: list of potential alternative solutions.

2.1.5 Determination of Criteria

Decision criteria, which will discriminate among alternatives, must be based on the goals. It is necessary to define discriminating criteria to measure how well each alternative achieves the goals. Since the goals will be represented in the form of criteria, every goal must generate at least one criterion but complex goals may be represented only by several criteria. If a goal does not suggest a criterion, it should be abandoned. Each criterion should measure something important, and not depend on another criterion. Criteria must discriminate among alternatives in a meaningful way.

It can be helpful to group together criteria into a series of sets that relate to separate and distinguishable components of the overall objective for the decision. This is particularly helpful if the emerging decision structure contains a relatively large number of criteria. Grouping criteria can help the process of checking whether the set of criteria selected is appropriate to the problem, can ease the process of calculating criteria weights in some methods, and can facilitate the emergence of higher level views of the issues. It is a usual way to arrange the groups of criteria, sub-criteria, and sub-sub-criteria in a tree-structure.

According to Baker et al. (2002), criteria should be

- Able to compare the performance of the alternatives.
- Complete to include all goals.

- Operational and meaningful.
- Non-redundant.
- Few in number.

Usually no one alternative will be the best for all goals, requiring alternatives to be compared with each other. The best alternative will be the one that most nearly achieves the goals.

Result: List of criteria representing the goals; collected criteria data for each alternative.

2.1.6 Evaluation of Alternatives Against Criteria

Alternatives can be evaluated with quantitative methods, qualitative methods, or any combination. Criteria can be weighted and used to rank the alternatives. Both sensitivity and uncertainty analyses can be used to improve the quality of the selection process. Experienced analysts can provide the necessary thorough understanding of the mechanics of the chosen decision-making method.

Every correct method for decision support needs, as input data, the evaluation of the alternatives against the criteria. Depending on the criterion, the assessment may be objective, with respect to some commonly shared and understood scale of measurement (e.g. money) or can be subjective (judgmental), reflecting the subjective assessment of the evaluator. After the evaluations the selected decision making tool can be applied to rank the alternatives or to choose a subset of the most promising alternatives.

Result: list of alternatives with defined measures of effectiveness.

2.1.7 Validation of Solution

After the evaluation process has selected a preferred alternative, the solution should be validated to ensure that it is able to solve the problem identified. It may happen that the decision making tool was misapplied. The comparison of the original problem statement to the goals and requirements is performed. A final solution should fulfill the desired state, meet requirements, and best achieve the goals. In complex problems the selected alternatives may also require for further goals or requirements modification and addition them to the decision model.

Once the preferred alternative has been validated, it can be presented as the final decision. A final result could report the decision process, assumptions, methods, and conclusions recommending the final solution.

2.2 Decision Support Methods

Decision support techniques are rational processes/systematic procedures for applying critical thinking to information, data, and experience in order to make a balanced decision when the choice between alternatives is unclear. They provide organized ways of applying critical thinking skills developed around accumulating answers to questions about the problem. Steps include clarifying purpose, evaluating alternatives, assessing risks and benefits, and making a decision. These steps usually involve *scoring* criteria and alternatives. This scoring (a systematic method for handling and communicating information) provides a common language and approach that removes decision making from the realm of personal preference or idiosyncratic behavior.

Depending on type of information used and way of achieving result (decision-making) the design supporting methods can be distinguished on three major approaches: Algorithmic, Knowledge-based inductive reasoning, and Case-based reasoning. First approach relies on specific procedure (algorithm, model) that transforms input to certain output; second method deals with generalised domain knowledge to make a decision; third one considers exemplary knowledge of designs.

2.2.1 Algorithmic Approach

The algorithmic design approach views the design process as the execution of an effective domain-specific procedure that yields a satisfying design solution in a finite number of steps. The main premise of this approach is that the initial requirements are well-defined and there are precisely defined criteria for determining whether or not an algorithm meets the requirements.

There exist a number of techniques which serve to optimize complex systems: exhaustive search, rapid search, mathematical programming. The search techniques involve many search strategies, such as breath-first, greedy methods, branch and bounds, dynamic programming and so on (Siddal, 1982; Dasgupta, 1989; Chandrasekaran, 1990). An exhaustive search generates an enormous number of alternatives to be considered, therefore the application of such techniques is limited. Search algorithms are judged on the basic of completeness, optimality, time complexity and space complexity. Complexity depends on the branching factor in the state space, and the depth of the shallowest solution. The alternative to an exhaustive search is rapid search, where a set of simple but arbitrary guidelines are adopted to limit the search space. The greatest disadvantage of any rapid search method is that the best solution might be out of the search space.

Mathematical programming techniques can be used to identify the potential design configuration based on the functional requirements. In general, in these methods the solution to the problem is developed by solving a mathematical model consisting of an objective function that is to be optimized and a set of constraints representing the limitation of the resources (Siddall, 1982; Braha and Maimon, 1998; Gani, 2004).

In chemical engineering design mathematical programming techniques are widely used. One of the targets in any industrial process design is to maximize the process-to-process heat recovery and to minimize the utility (energy) requirements. This goal can be achieved by utilizing Pinch Technology. This technique presents a simple methodology for systematically analysing chemical processes and the surrounding utility systems with the help of the First and Second Laws of Thermodynamics. Pinch Analysis is used to identify energy cost and heat exchanger network (HEN) capital cost targets for a process and recognizing the pinch point (Townsend and Linnhoff, 1983).

Another method utilised for process synthesis is the superstructure generation with following optimization (Grossmann, 1985). The advantage of the approach is the rigorous analysis of features such as structure interactions and capital costs. The disadvantage of the method is the need for a big computational efforts and the fact that the optimality of the solution can only be guaranteed among alternatives considered a priori.

An incomplete, ill-structured design problem may be decomposed into one or more well-structured components, and then the algorithmic methods may be successfully utilized to solve each of these well-structured sub-problems.

2.2.2 Knowledge-Based Inductive Reasoning Approach

This approach to decision support is based on capturing knowledge of a certain domain and using it to solve problems. The design is considered as a problem-solving process of searching through a state-space, from initial problem state to the goal state. Transition from one step to another is affected by applying one of a finite set of operators, based on functional requirements and design constraints (domain specific knowledge) and meta-rules (domain independent knowledge).

Due to emphasis of knowledge, such computer systems are known as knowledge-based or expert systems. The term 'expert system' is often used as the input knowledge is usually acquired from human experts. When knowledge is generally acquired through non-human intervention (computer methods), the term 'knowledge-based system (KBS) is more appropriate. The united term 'knowledge-based expert system' (KBES) is further used to represent both or combined methods of knowledge acquisition.

KBES is able to use previously defined rules to solve a new problem. Inductive reasoning, implemented in KBES, means reaching conclusions about a whole class of facts based on evidence on part of that class. KBESs are examples of automatic problem-solvers that rely on domain-specific heuristics.

Such reasoning differs from algorithmic approach with following issues:

- Simulation of human reasoning about a problem domain, rather than modelling the domain itself;
- Reasoning over representation of human knowledge, in additional to doing numerical calculation or data retrieval;
- Suggesting a solution to a problem using heuristic or approximate methods which, unlike to algorithmic solution, are not guaranteed to succeed;
- Capability to explain and justify solutions or recommendations to convince that the reasoning result is correct.

Algorithmic approach is the reasoning strategy which is guaranteed to find the solution to whatever the problem is, if there is such a solution. For the large, difficult problems with which expert systems are concerned, it may be more useful to employ heuristics: strategies that often lead to the correct solution, but which also sometimes fail. Humans use heuristics in their problem solving. If the heuristic does fail, it is necessary for the problem solver to either pick another heuristic, or know that it is appropriate to give up. In design problems, there may be many millions of possible solutions to the problem as presented. It is not possible to consider each one in turn, to find the right (or best) solution; heuristically-guided search is required.

Some rules used for inductive reasoning in KBES may only express a probability that a conclusion follows from certain premises, rather than a certainty. The items in the knowledge base must reflect this uncertainty, and the inference engine must process the uncertainties to give conclusions that are accompanied by likelihood that they are true. Assumptions – for instance, about the reliability of a piece of evidence – may have to be abandoned part way through the reasoning process.

Expert systems usually contain inference engine, knowledge base and two interfaces to communicate with user and experts (Fig. 2.1). Knowledge based system instead of expert interface includes knowledge generation part.

The inference engine is responsible for extracting appropriate rules from knowledge base and generating new information. There are two main ways for inference: forward chaining and backward chaining. The forward chaining is used for problem-solving when data obtained from communication with the user are the starting point. The system attempts to achieve conclusions. A problem with forward chaining is that many goals are possible to achieve whether useful or not. In contrast, backward chaining, often described as

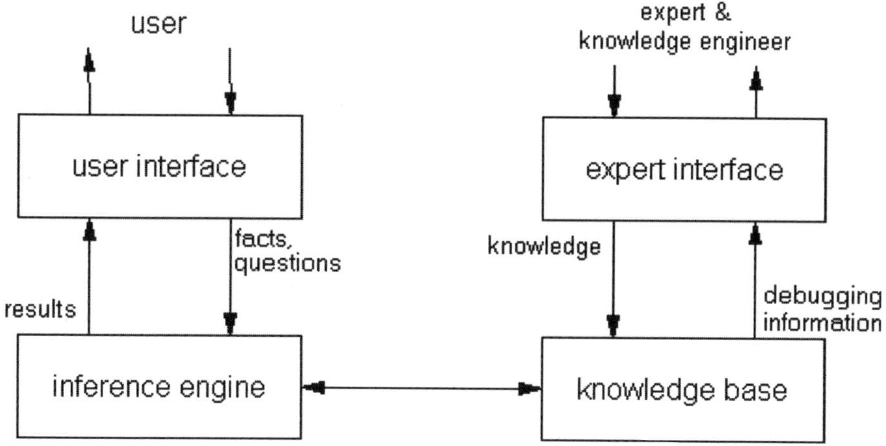

Fig. 2.1. Expert system layout

goal-directed reasoning, starts with a hypothesis or specific goal and then attempts to find data from interaction with the use to prove or disprove the conclusion. Whereas the forward chaining is often used in KBES developed for design problems, backward chaining is specifically applicable to troubleshooting and control problems. These methods of inference can often be combined in KBES.

An inference engine may also have the capability to reason in the presence of uncertainty both in the input data and also in the knowledge base. The major methods are Bayesian probabilities and fuzzy logic.

KBES approach is the base for many of the computer-aided design systems developed in recent years (Tong and Sriram, 1992; Wilke et al., 1998; Nakayama and Tanaka, 1999). A review of knowledge-based methods for design tasks in chemical engineering has recently been presented by Li and Kraslawski et al. (2004).

The knowledge-based inductive reasoning approach is very useful for solving tightly coupled, highly integrated design problems. However, when faced with an original design problem with no previous rules to help it, expert systems are incapable of original creativity.

Knowledge-based expert systems are based upon an explicit model of the knowledge required to solve a problem – so called second generation systems (Clancey, 1985) using a deep causal model that enables a system to reason using first principles. But whether the knowledge is shallow or deep an explicit model of the domain must still be elicited and implemented often in the form of rules or perhaps more recently as object models. The tight problem of KBES in many sectors is knowledge acquisition, often being referred to as the knowledge elicitation bottleneck. To overcome this difficulty special information techniques can be applied. The knowledge can be collected with decision tree generated by various algorithms. Despite obvious advantages automotive generation of knowledge base (decision tree) has several difficulties:

- Only classification problems can be addressed.
- Human interventions are still required to define attributes and original knowledge matrix.
- When new examples become available it is necessary to rebuild the existing tree.

An expert system is purposed to perform at a human expert level in a narrow, specialised domain. Thus, the most important characteristic of KBES is its high-quality performance. A unique feature of an expert system is its explanation capability. This enables the KBES to review its own reasoning and explain its decisions. An explanation in expert system in effect traces the rules fired during a problem-solving session.

KBES employs symbolic reasoning when solving a problem. Symbols are used to represent different types of knowledge. Algorithmic approach always performs the same operations in the same order, and it always provide an exact solution (if it is principally possible). Unlike algorithmic approach, KBES do not follow a prescribed sequence of steps. It permits inexact reasoning and can deal with incomplete, uncertain and fuzzy data.

2.2.3 Case-Based Reasoning Approach

Case-based problem solving is based on the premise that a design problem solver makes use of experiences (cases) in solving new problems instead of solving every new problem from scratch (Kolonder, 1993). Lansdown (1987) argues that "innovation arises from incremental modification of existing ideas rather than entirely new approaches". Coyne et al. (1990) classify the case-based approach into three activities: creation, modification, and adaptation. Creation is concerned with incorporating requirements to create a new prototype. Modification is concerned with developing a working design from a particular category of cases. Adaptation is concerned with extending the boundaries of the class of the cases.

Case-based reasoning (CBR) solves new problems by adapting previously successful solutions to similar problems. It has several features, which make this approach different from KBES, namely:

- CBR does not require an explicit domain model and elicitation becomes a task of gathering case histories.
- Implementation is reduced to identifying significant features that describe a case, an easier task than creating an explicit model.
- Largely volumes of information can be managed.
- CBR systems can learn by acquiring new knowledge as cases thus making maintenance easier.

A case-based reasoning approach can handle incomplete data: it is robust with respect to unknown values because it does not generalize the data. Instead, the approach supports decision making relying on particular experiences.

2.3 Knowledge Engineering

The described above approaches to decision support in design deal with knowledge of certain organization. Different approaches have different knowledge organizations. However, the process of acquisition, structuring and representation of knowledge precedes any reasoning activity and it can be regarded as common for all approaches. This process is known as knowledge engineering.

There are two main views to knowledge engineering. The traditional view is known as "Transfer View". In this view, the assumption is to apply conventional knowledge engineering techniques to transfer human knowledge into artificial intelligent systems. The alternative view is known as the "Modeling View". In this view, the knowledge engineer attempts to model the knowledge and problem solving techniques of the domain expert into the artificial intelligent system.

Knowledge engineering relates to the building, maintaining and development of knowledge-based systems. It has a great deal in common with software engineering, and is related to many computer science domains such as artificial intelligence, databases, data mining, expert systems, and decision support systems.

Various activities of KE specific for the development of a knowledge-based system:

(1) Assessment of the problem
(2) Development of a knowledge structure
(3) Implementation of the structured knowledge into knowledge-bases
(4) Acquisition and structuring of the related information, knowledge and specific preferences
(5) Testing and validation of the inserted knowledge
(6) Integration and maintenance of the system
(7) Revision and evaluation of the system.

KE deals with the knowledge, and mainly with the structure (organization) of knowledge. Therefore, the organisation of knowledge is a key element of KE.

2.3.1 Classification of Knowledge

Initial source of knowledge base is a set of data. Data refers to facts, codes, marks and signals. Data is transformed by processing to information which is organized to be meaningful to the object receiving it. Knowledge can therefore be regarded as information which is understood and can be applied to get new information (Fig. 2.2).

Knowledge can be derived from other knowledge. Priori perceived knowledge can be transcribed to five primary types of content: facts, concepts, processes, procedures, and principles (Clark and Chopeta, 2004).

Facts are specific and unique data or instance. Concept is a class of items, words, or ideas. There are two types of concepts: concrete and abstract. Process is represented by a flow of events or activities that describe how things work rather than how to do things. There are normally two types: business processes that describe work flows and technical processes that describe how things work in equipment or nature. Procedures are series of step-by-step actions and decisions that result in the achievement of a task. There are two types of actions: linear and branched. Guidelines and rules form principles. It includes not only what should be done, but also what should not be done. Principles allow one to make predictions and draw implications. Given an effect, one can infer the cause of phenomena. Principles are the basic building blocks of causal models or theoretical models (theories).

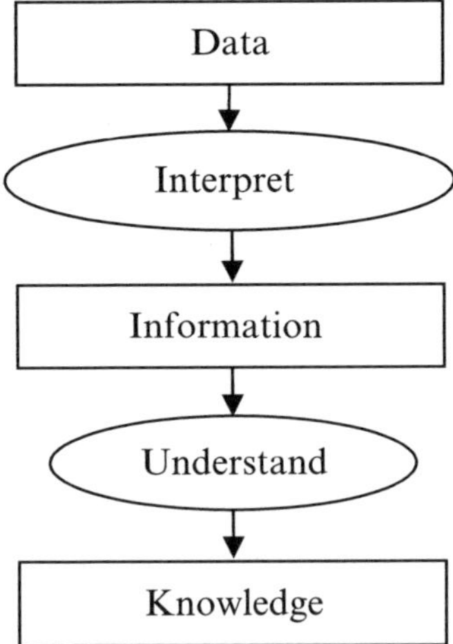

Fig. 2.2. Transformation data to knowledge

These contents can be used to create two categories of knowledge: declarative and procedural, where the first comprises concepts and the second are actions.

Declarative Knowledge

Declarative knowledge refers to representations of objects and events and how these knowledge and events are related to other objects and events. They focus on the why rather than the how. Declarative models include propositions and schemata. Proposition consists of a predicate or relationship and at least one argument. Schemata are higher-level cognitive units that use propositional networks as their building blocks. These are often abstract or general in nature that allows to classify objects or events as belonging to a particular class and to reason about them. Schemata can be conceptional knowledge, plan-like knowledge, and causal knowledge.

Concepts are simple schemata that represent a class of objects, events, or other entities by their characteristic features. Concepts enable a person to identify or classify particular instances (concrete object or event) as belonging to a particular class. In a language, most words identify concepts and at least to a certain degree, they are arbitrary in that they can be categorized in many alternative ways.

Experts possess more powerful concepts in their domain than novices that help them to solve problems. These concepts give them patterns for labeling various memory states, which allow them to classify problems according to their solution mode or deep structure. Where as novices typically classify problems according to their surface structure or superficial feature.

Plan-Like Knowledge is simple schemata that describe how goals are related in time or space. They allow us to understand events and organize functions and actions. Plans are often referred to as scripts (or simple procedures) because they represent routine sequences of events.

Causal Knowledge is complex schemata that link principles and concepts with each other to form cause–effect relationships. They are able to interpret events, give explanations, and make predictions.

Procedural Knowledge

Procedural models focus on tasks that must be performed to reach a particular objective or goal. It is characterized as knowing how. Procedural knowledge is often difficult to verbalize and articulate (tacit knowledge) than declarative knowledge.

Procedural knowledge emphasizes hierarchical or information processing approaches based upon productions. A combination of productions creates production systems. Productions are the building blocks of procedural knowledge and are composed of a condition and an action or IF and THEN statement. A production system is a set of productions for cognitive processing. It is characterized by the recognize-act cycle in which one production leads to another production. There are two types of productions: rules and heuristics. The difference between rules and heuristics is based on the validity and rigour of the arguments used to justify them – rules are always true, valid and can be justified by arguments; heuristics are the expert's best judgments, may not be valid in all cases and can only be justified by examples.

There also can be distinguished the specific class of knowledge, which stays above of previous declared categories of knowledge, called meta-knowledge.

Meta-Knowledge

Meta-knowledge is knowledge about knowledge. More precisely speaking, meta-knowledge is systemic problem and domain-independent knowledge which performs or enables operations on another more or less specific domain-dependent knowledge in different domains/areas of human activities. Meta-knowledge is a fundamental conceptual instrument in such research and scientific domains as, knowledge engineering, knowledge management, and others dealing with study and operations on knowledge, seen as an unified object/entities, abstracted from local conceptualizations and terminologies.

Examples of the first-level individual meta-knowledge are methods of planning, modeling, learning and every modification of a domain knowledge. The procedures, methodologies and strategies of teaching, coordination of e-learning courses are individual meta-meta-knowledge of an intelligent entity (a person, organization or society). The universal meta-knowledge frameworks have to be valid for the organization of meta-levels of individual meta-knowledge.

Knowledge can be classified according to the origin of the knowledge. The source of *empirical* knowledge is practical experience. Observations are made when running the process. *Theoretical* knowledge is based on natural laws and scientific theories. The third form of knowledge is subjective, *experience-based* knowledge.

When describing certain domain, general and problem independent knowledge is called *background* knowledge. If the background knowledge describes a specific part of the domain it is called *contextual* knowledge. *Episodic* knowledge is of narrative character. It records the story of something happened in the past.

There are two levels of knowledge: shallow or deep knowledge. *Shallow* knowledge can deal with very specific situations, whereas the *deep* knowledge is a representation of all information of a domain.

2.3.2 Knowledge Acquisition

The objective of knowledge acquisition is to collect or elicit knowledge from the experts and other sources and structure it in a certain way.

The first step of knowledge acquisition is to collect all the potential sources of knowledge. They are text book written specifically in the domain, research and technical reports, journal articles, reference manuals, case studies, operational procedures and organizational policy statements. Availability of documents may vary; in some domain there may be many available, and in others none at all. The reports and books contain factual knowledge; they are often detailed, precise and well structured but are not always relevant to knowledge acquisition task. Often, the analysis of significant amount of documents is highly time-consuming. The range of problems which textbooks examine and solve is always smaller than the range of problems that a human expert is master of.

Knowledge can also be obtained from discussion with organization personnel like projects leader and consultants. The most important branch of knowledge acquisition is knowledge elicitation – obtaining knowledge from domain experts.

Expert knowledge includes:

- Domain-related facts and principles
- Modes of reasoning
- Reasoning strategies
- Explanations

Two kinds of knowledge can be elicited from experts:

- Explicit knowledge is the knowledge which the domain expert is able to articulate.
- Tacit knowledge is the knowledge which the domain expert is not conscious of having but does exists as proved by expert's known capability of solving problems in the domain.

Explicit knowledge is easy to elicit from experts since it is mainly factual in nature. Tacit knowledge is difficult to identify and elicit but it is essential for successful development of knowledge-based systems. Knowledge obtained from experts have following features: incomplete – experts may forgot, superficial – exerts often cannot go to details, imprecise – experts may not know exact detail, inconsistent – when expert fall into contradictions, incorrect – when experts may be wrong.

Such features could rise a lot of problems in creation of knowledge base. Needs in communication with experts as well as in retrieval data from various documents exists more or less in all approached for decision making support. But the acquired knowledge have to be interpreted and translated into the rules and heuristics in the KBES approach, which is also time-demanding task. In contrast to, CBR approach relies only on set of acquired information (even not knowledge in many cases).

In addition to manual methods of knowledge acquisition there are automated methods whereby the computers are used. Using a computer for a knowledge acquisition overlaps with software engineering problems.

2.3.3 Software Engineering versus Knowledge Engineering

Software engineering provides the mechanisms for validating the implementation of well specified algorithms. Human–computer interaction provides analysis and design techniques based on prototyping of the user interface to address aspects of systems where the risks are associated with the users' needs, or the system usability. Data engineering addresses the permanent storage of large amounts of data and the efficient retrieval of the relatively small portion required for any process. In contrast, knowledge engineering addresses the structure of complex but ill-defined processes where the solution to defining the process is to define the knowledge involved in the process explicitly in a knowledge-based system (KBS).

Conventional software development follows the waterfall life cycle model. This requires complete system requirements at the start of development. Errors later in development can be fixed at little cost; errors at the start of development incur large costs. If the risks of failure of the project are associated with the efficiency of the implementation of a system this is appropriate. If the risk of a project failing is due to the uncertainty of the algorithms to perform the functions required, user requirements or enterprise objectives then an approach which is flexible at the start of the process is appropriate.

Conventional software engineering approaches produce efficiently implemented code to execute algorithms to perform required functions which will always produce the correct outcome for correct input. The knowledge engineering approach allows users and experts to describe requirements and methods to perform the required functions at a high level close to the one in which they think about the task: the Knowledge Level. These can then be presented back to them for validation of the content, and modification.

If algorithms to perform the required functions cannot be determined then heuristics which produce correct outcomes sufficiently often for some task requirements can be used – there may not be sufficiently detailed domain theory to supply algorithms so human expertise in the domain can be used.

If heuristic knowledge cannot be acquired which produces correct outcomes sufficiently frequently then the project should be terminated – there may not be domain expertise to acquire. Since this possibility continues after initial problem definition (including feasibility studies) into the acquisition of knowledge, then staged contracting should be used to protect the client, and the commitments made by the developer.

Knowledge engineering differs from conventional software engineering mainly at the early stages of the life cycle when user requirements and functional methods (or knowledge) are being acquired. The tools for implementation, user interface design, testing, maintenance and updating systems may differ, but the principles which govern all software systems are the same. Therefore, although the early stages of knowledge acquisition will involve a knowledge engineer and a (or more) domain experts, later stages will involve software engineers for implementation/integration.

2.3.4 Knowledge Representation

Knowledge representation (KR) is the study of how knowledge about the world can be represented and what kinds of reasoning can be done with that knowledge. Important questions include the tradeoffs between representational adequacy, fidelity, and computational cost, how to make plans and construct explanations in dynamic environments, and how best to represent default and probabilistic information

A variety of ways of representing knowledge in a knowledge base have been developed over the years.

The commonly used methods for knowledge representation are production rules, frames, semantic networks, ontology and objects.

Production Rules

They express the relationship between several pieces of information. The rules are conditional statements that specify actions to be taken or advice to be followed under certain sets of conditions.

Each production rule implements an autonomous piece of knowledge and can be developed and modified independently of other rules. However, when combined, a set of rules may yield better results that the sum of results of the individual rules and independency is lost. It must be taken into account when adding new rules to a current knowledge base to avoid conflicts.

Frames

They are templates for holding clusters of related knowledge about a particular object. They are able to represent the attribute of an object in a more descriptive way that is possible using production rules. The frame typically consists of a number of slots which, like attributes, may or not contain a value.

Semantic Network

Because any knowledge incorporates concepts and will be expressed using terms, the interdependencies between knowledge and language are essential for the definition itself.

A semantic network is a directed graph consisting of vertices, which represent concepts, and edges, which represent semantic relations between the concepts. Such networks involve fairly loose semantic associations that are nonetheless useful for human browsing. It is possible to represent logical descriptions using semantic networks such as the existential graphs or the related conceptual graphs. These have expressive power equal to or exceeding standard first-order predicate logic. The semantic networks can be used for reliable automated logical deduction. Some automated reasoners exploit the graph-theoretic features of the networks during processing.

One can consider a mind map to be a very free form variant of a semantic network. By using colors and pictures the emphasis is on generating a semantic net which evokes human creativity. However, a fairly major difference between mind maps and semantic networks is that the structure of a mind map, with nodes propagating from a centre and sub-nodes propagating from nodes, is hierarchical, whereas semantic networks, where any node can be connected to any node, have a more heterarchical structure.

Ontology

An ontology is a knowledge model that represents a set of concepts within a domain and the relationships between those concepts. The word ontology means “the study of the state of being”. An ontology describes the states of being of a particular set of things. This description is usually made up of axioms that define each thing. It is used to reason about the objects within that domain.

Ontologies generally describe:

- *Individuals*: the basic objects
- *Classes*: sets, collections, or types of objects
- *Attributes*: properties, features, characteristics, or parameters that objects can have and share
- *Relations*: ways that objects can be related to one another
- *Events*: the changing of attributes or relations

The individuals in an ontology may include concrete objects such as tables, automobiles, molecules, and reactor, as well as abstract individuals such as numbers and words. Actually, an ontology need not include any individuals, but one of the general purposes of an ontology is to provide a means of classifying individuals, even if those individuals are not explicitly part of the ontology.

Classes may contain individuals, other classes, or a combination of both. Ontologies vary on whether classes can contain other classes, whether a class can belong to itself, whether there is a universal class (that is, a class containing everything), etc. The classes of an ontology may be extensional or intensional in nature. A class is extensional if and only if it is characterized solely by its membership. If a class does not satisfy this condition, then it is intensional. While extensional classes are more well-behaved and well-understood mathematically, they do not permit the fine grained distinctions that ontologies often need to make.

A partition is a set of related classes and associated rules that allow objects to be placed into the appropriate class. If the partition rules guarantee that an object cannot be in both classes, then the partition is called a *disjoint* partition. If the partition rules ensure that every concrete object in the superclass is an instance of at least one of the partition classes, then the partition is called an *exhaustive* partition.

Objects in the ontology can be described by assigning attributes to them. Each attribute has at least a name and a value, and is used to store information that is specific to the object it is attached to. The value of an attribute can be a complex data type.

An important use of attributes is to describe the relationships between objects in the ontology. Typically a relation is an attribute whose value is another object in the ontology. The most important type of relation is the subsumption relation (knows as *is-a*). This defines which objects are members of classes of objects.

The addition of the *is-a* relationships has created a hierarchical taxonomy; a tree-like structure that clearly depicts how objects relate to one another. Another common type of relations is the Meronymy relation (written as *part-of*) that represents how objects combine together to form composite objects. The examples of described relation types are represented in Fig. 2.3.

As well as the standard is-a and part-of relations, ontologies often include additional types of relation that further refine the semantics they model. These relations are often domain-specific and are used to answer particular types of question.

Knowledge Representation Languages and Ontology Analysis

One of the developments in the application of KR has been the proposal (Minsky, 1981) and development (Brachman and Schmolze, 1985) of frame-based KR languages. While frame-based KR languages differ in varying

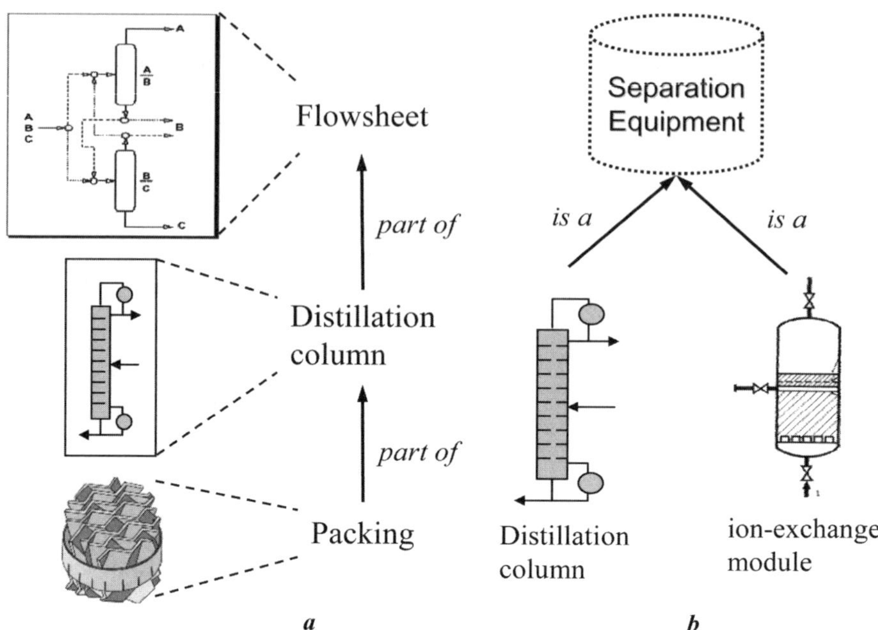

Fig. 2.3. Meronymy (**a**) and subsumption (**b**) relations examples

degrees from each other, the central tenet of these systems is a notation based on the specification of objects (concepts) and their relationships to each other. The main features of such a language are:

- Object-orientedness. All the information about a specific concept is stored with that concept, as opposed, for example, to rule-based systems where information about one concept may be scattered throughout the rule base.
- Generalization/Specialization. Long recognized as a key aspect of human cognition (Minsky, 1981), KR languages provide a natural way to group concepts in hierarchies in which higher level concepts represent more general, shared attributes of the concepts below.
- Reasoning. The ability to state in a formal way that the existence of some piece of knowledge implies the existence of some other, previously unknown piece of knowledge, is important to KR. Each KR language provides a different approach to reasoning.
- Classification. Given an abstract description of a concept, most KR languages provide the ability to determine if a concept fits that description, this is actually a common special form of reasoning.

Object orientation and generalization help to make the represented knowledge more understandable to humans, reasoning and classification help make a system behave as if it knows what is represented.

It is important to realize both the capabilities and limitations of frame-based representations, especially as compared to other formalisms. To begin with, all symbolic KR techniques are derived in one way or another from First Order Logic, and as a result are suited for representing knowledge that does not change. Different KR systems may be able to deal with non-monotonic changes in the knowledge being represented, but the basic assumption has been that change, if present, is the exception rather than the rule.

Two other major declarative KR formalisms are production systems and database systems. Production systems allow for the simple and natural expression of IF-THEN rules. However, these systems have been shown to be quite restrictive when applied to large problems, as there is no ordering of the rules, and inferences cannot be constrained away from those dealing only with the objects of interest. Production systems are subsumed by frame-based systems, which additionally provide natural inference capabilities like classification and inheritance, as well as knowledge-structuring techniques such as generalization and object orientation.

Database systems provide only for the representation of simple assertions, without inference. Rules of inference are important pieces of knowledge about a domain.

What makes up a specific domain ontology is restricted by the representational capabilities of the meta-model – the language used to construct the model. Each knowledge representation language differs in its manner and range of expression. In general, an ontology consists of three parts: concept definitions, role definitions, and further inference definitions.

The concept definitions set up all the types of objects in the domain. In object oriented terms this is called the class definitions, and in database terms these are the entities. There can be three parts to the concept definitions: concept taxonomy, role defaults and role restrictions. The taxonomy is common to most knowledge representation languages, and through it is specified the nature of the categories in terms of generalization and specialization. Role defaults specify for each concept what the default values are for any attributes. Role restrictions determine for a concept any constraints on the values in a role, such as what types the values must be, how many values there can be, etc.

A role is an attribute of an object. In object-oriented terms it is a slot, in database terms (and even some KR languages) it is a relation. Roles which represent relationships are unidirectional. A role definition may have up to three parts as well: the role taxonomy which specifies the generalization/specialization relationship between roles; the role inverses which provide a form of inference that allows the addition of a role in the opposite direction when the forward link is made; and the role restrictions where the role itself may be defined such that it can only appear between objects of certain types (domain/range restrictions), or can only appear a specified number of times (cardinality restriction).

The final part of an ontology is the specification of additional inference that the language provides. Examples of this are forward and/or backward chaining rules, path grammars, subsumption and/or classification, demons, etc.

Knowledge Engineering must address the issue of reliable methodology to meet the practical engineering objectives it now has. Secondly, the systems produced through knowledge engineering methods must be able to re-use not only abstract ideas, but also implementation level knowledge. To do these issues of portability and interoperability must be addressed. A consequence of addressing these two issues could be to lose the apparent freedom provided by expert systems and to become bound by the formalities of software engineering. To avoid this, knowledge engineering must maintain its influence on user interfaces and the ability of KBS to explain their reasoning.

2.4 Decision Supporting Systems

Decision making in design often requires access to and the processing of a large amount of data and logical relations which (due to the nature of the problem) cannot or should not be replaced by the intuition of decision maker. In many design situations it is not a small task to examine even the possible range of feasible alternatives. In the context of decision support, the problem is a situation description in which information is missing. The goal is to complete the situation description until the demand for information is satisfied. The use of computers for processing situations leads to implementing a Decision Supporting System.

A Decision Supporting System (DSS) is a supportive tool for the management and the processing of large amounts of information and logical relations that helps a decision maker (design engineer) to extend his vision and thus help to reach a better decision. In other words, a DSS can be considered as a tool that performs the task of data processing and provides relevant information that enables a design engineer to concentrate on the part of the decision making process that cannot be formalized.

Because there are many approaches to decision-making and because of the wide range of domains in which decisions are made, the concept of DSS is very broad. A DSS can take many different forms. In general, a DSS is an information system that provides the ability to analyze information and predict the impact of decisions before they are made. A decision is a choice between alternatives based on estimates of the values of those alternatives. Supporting a decision means helping people working alone or in a group gather intelligence, generate alternatives and make choices. Supporting the choice making process involves supporting the estimation, the evaluation and/or the comparison of alternatives. In practice, references to DSS are usually references to computer applications that perform such a supporting role.

The goal of a DSS is to supplement the decision powers of the human with the data manipulating capabilities of the computer (Emery, 1987). It is not intended to solve a decision problem. Therefore it should not support reaching a single or unique decision nor should it restrict a possible range of decisions.

Furthermore, it is usually not possible to decide whether a solution found by a DSS is correct or not. Rather, this information may be more or less useful; it may be better or worse than other information (Lenz et al., 1998).

Richter (1992) identified four characteristic properties of DSS:

(1) The amount of information that has to be coped with is too large to be handled by humans without the support of an appropriate technical system.
(2) The decision has to be made quickly.
(3) Data has to be prepared for decision making.
(4) The process of decision making is highly complex and requires specific algorithms.

Turban et al. (2005) composed more longer list of ideal characteristics and capabilities of DSS:

1. Support for decision makers in semistructured and unstructured problems.
2. Support managers at all levels.
3. Support individuals and groups.
4. Support for interdependent or sequential decisions.
5. Support intelligence, design, choice, and implementation.
6. Support variety of decision processes and styles.
7. DSS should be adaptable and flexible.
8. DSS should be interactive and provide ease of use.
9. Effectiveness balanced with efficiency (benefit must exceed cost).
10. Complete control by decision-makers.
11. Ease of development (modification to suit needs and changing environment).
12. Support modeling and analysis.
13. Data access.
14. Standalone, integration and Web-based.

2.4.1 Classification of DSS

There is no universally accepted classification of DSS. Different authors propose different classifications. Using the relationship with the user as the criterion, Häettenschwiler (1999) differentiates *passive*, *active*, and *cooperative* DSS. A passive DSS is a system that aids the process of decision making, but that cannot bring out explicit decision suggestions or solutions. An active DSS can bring out such decision suggestions or solutions. A cooperative DSS allows the decision maker (or its advisor) to modify, complete, or refine the decision suggestions provided by the system, before sending them back to the system for validation. The system again improves, completes, and refines the suggestions of the decision maker and performs the validation. The whole process then starts again, until a consolidated solution is generated.

Using the mode of assistance as the criterion, Power (2002) differentiates *communication-driven* DSS, *data-driven* DSS, *document-driven* DSS, *knowledge-driven* DSS, and *model-driven* DSS.

A model-driven DSS emphasizes access to and manipulation of a statistical, financial, optimization, or simulation model. Model-driven DSS use data and parameters provided by users to assist decision makers in analyzing a situation; they are not necessarily data intensive. Dicodess is an example of an open source model-driven DSS generator (Gachet, 2004).

A communication-driven DSS supports more than one person working on a shared task; examples include integrated tools like Microsoft's NetMeeting or Groove (Stanhope, 2002).

A data-driven DSS or data-oriented DSS emphasizes access to and manipulation of a time series of internal company data and, sometimes, external data.

A document-driven DSS manages, retrieves and manipulates unstructured information in a variety of electronic formats.

A knowledge-driven DSS provides specialized problem solving expertise stored as facts, rules, procedures, or in similar structures.

Using scope as the criterion, Power (1997) differentiates *enterprise-wide* DSS and *desktop* DSS. An enterprise-wide DSS is linked to large data warehouses and serves many managers in the company. A desktop, single-user DSS is a small system that runs on an individual personal computer.

2.4.2 Architectures of DSS

Different authors identify different components in a DSS. Sprague and Carlson (1982) identify three fundamental components of DSS:

- The database management system (DBMS)
- The model-base management system (MBMS)
- The dialog generation and management system (DGMS)

Haag et al. (2006) describe these three components in more detail: the DBMS stores data, which can be further divided into that derived from the local data repositories, from external sources such as the Internet, or from the personal insights and experiences of individual users; the MBMS handles representations of events, facts, or situations using various kinds of models; and the DGMS is the component that allows a user to interact with the system.

According to Power (2002), academics and practitioners have discussed building DSS in terms of four major components: the user interface, the database, the model and analytical tools, and the DSS network.

Häettenschwiler (1999) identifies five components of DSS:

- The users with different roles or functions in the decision making process (decision maker, advisors, domain experts, system experts, data collectors)
- The specific and definable decision context

- The target system describing the majority of the preferences
- The knowledge base made of external data sources, knowledge databases, working databases, data warehouses and meta-databases, mathematical models and methods, procedures, inference and search engines, administrative programs, and reporting systems, and
- The working environment for the preparation, analysis, and documentation of decision alternatives

Marakas (1999) proposes a generalized architecture made of five distinct parts:

- The data management system
- The model management system
- The knowledge engine
- The user interface, and
- The user(s)

Holsapple and Whinston (1996) classify DSS into the following six frameworks: Text-oriented DSS, Database-oriented DSS, Spreadsheet-oriented DSS, Solver-oriented DSS, Rule-oriented DSS, and Compound DSS.

The support given by DSS can be separated into three distinct. interrelated categories (Hackathorn and Keen, 1981): Personal Support, Group Support and Organizational Support.

DSSs which perform selected cognitive decision-making functions and are based on artificial intelligence or intelligent agents technologies are called Intelligent Decision Support Systems (IDSS).

A DSS is a problem dedicated system usually designed for a specific decision making process and its environment. Using DSS is useful in complex design situations for which specification of attainable goals and rational decisions is quite complicated. The DSS finds a solution closest to the specified goals. This ability to provide answers for decision support in a changing environment is the main advantage of decision supporting systems.

There are two alternative approaches for the design of DSSs: normative and descriptive (Lenz et al., 1998). The normative approach attempts to establish general rules for rational behaviour. It is realized by utilizing a knowledge-based reasoning technique. On other hand, the descriptive approach does not rely much on general principles but on examples of successful problem solving episodes. Such episodes are investigated to obtain knowledge about how the solution was derived. This can clearly be implemented by utilizing a case-based reasoning approach.

2.5 Conclusions

In the chemical process design there is a growing demand for an improvement to the design process in order to generate better flowsheets within a shorter development time. Existing design supporting tools have been developed for specific purposes and related to separate parts of process design. Therefore a tool or methodology that is able to support overall design activity (from A to F levels) would be very valuable.

Due to uncertain and incomplete input data and the lack of formal methods, approaches to innovative design and redesign support are proposed to assist the design engineer rather than to automate the process. Engineer intervention is required to generate or evaluate a proper solution. The problem solving process then is to provide the user of a design supporting system with documents to satisfy his demands.

Knowledge-based systems using rule-based reasoning and various algorithmic techniques have been applied to build design decision support system. Although such systems have been met with some success, difficulties have been encountered in terms of formalizing generalized design experiences as rules, logic and domain models. In order to support innovative design tasks, conventional problem solving methods are not applicable, in general. The use of experience is of particular importance. Recently, researchers have been exploring the idea of using case-based reasoning to complement or replace other approaches to design support. In order to support creative design tasks, the application of analogical problem solving is advantageous.

The idea of supporting the designer by means of case-based knowledge to help navigate through a dynamic design process seems to be promising. Moreover, a general approach which can support various stages of design activity is only possible with case-based reasoning: it relies on particular experience of design and there is no need for derivation of specific heuristics of the design process for each design stage.

Case-based reasoning (CBR) can support innovative design and redesign activity by reminding designers of previous experiences that could match with the new design situation, not necessary totally but only partially. This approach is able to support almost all steps of chemical process design, except perhaps the first and last ones (i.e. from B to E). But even for steps A and F, the sort of supporting activity can be realised. The next part describes a case-based design supporting paradigm.

Part II

Case-based Design Support

3

Case-Based Reasoning Approach

As stated in the previous part, design activity is based on a generic problem solving process which begins with problem definition and description, involves various forms of analysis, might include simulation and modelling, moves to generation of solutions and thence to the evaluation of alternative solutions to the problem. Decisions characterise every stage of this process. The process takes place across many levels of abstractions and is iterative in form. Meanwhile, practice shows that often it is more efficient to solve a new problem by starting with a solution of a previous, similar problem than to generate the entire solution from scratch. Whenever it is easier or more convenient to reuse experience, humans prefer to do so rather than to derive a completely novel solution. A design engineer may find it difficult to determine the similar elements between a new problem and massive historical data; moreover, the similarities are often not easily noticeable. In order to facilitate the design process and to reduce the required development time, a decision supporting system utilising the case-based reasoning approach could be used to support process and product design.

Case-based decision support systems try to help in decision making by providing cases similar to an actual problem situation. The cases are usually advice or a useful piece of information. In the context of decision support, the problem is a situation description in which information is missing. The goal is to complete the description of the situation until the demand for information is satisfied. Thus, the problem solving process is to provide the user with records to satisfy his demands appropriately. This chapter reviews the different models of the case-based reasoning and describes the case-based methodology to support decision making in design.

3.1 Case-Based Reasoning Concept

Case-based reasoning (CBR) is a problem solving technique. It imitates human thinking trying to make a decision based on earlier experiences. Its history starts in about the year 1977 and originated in cognitive science. The idea

Y. Avramenko and A. Kraslawski: *Case-Based Design*, Studies in Computational Intelligence (SCI) **87**, 51–70 (2008)
www.springerlink.com

of CBR can be formulated in one sentence: a Case-Based reasoner solves new problems by using or reusing solutions that were used to solve similar problems.

Kolonder (1993) defined the case-based reasoning as "adapting old solutions to meet new demands, using old cases to explain new situations, using old cases to critique new solutions, or reasoning from precedents to interpret a new situation". Case-based reasoning suggests a model of reasoning that incorporates problem solving, learning, and integrates all with memory processes.

Reference to previous similar situations is often necessary to deal with the complexity of novel situations. A new problem situation needing to be solved must be identified among other old pieces of experience which is recorded. It means finding in memory the experience closest to a new situation. Recalling cases is at the core of case-based reasoning. A retrieved case is subject to reuse. It means adapting the old case to modify an old solution in order to meet the requirements of a new situation. Thus, the key notions of CBR are the case, retrieval, reuse (adaptation).

3.1.1 Representation of Experience

An experience situation is represented by a case. It can be represented by a rule, a constraint, some general law, advice, or simply by recording a past event. The experience recorded in such a case thus reflects just a single event. Cases, which represent specific knowledge, may display how a task was carried out, or how a piece of knowledge was applied. The episode of experience recorded in a case has to contain some decisions which could be found useful.

In decision making applications, a case is considered as a problem-solving episode that contains a problem and solution. Thus, a single case is represented as a pair: a problem and its solution. Many cases are collected in a set to build a case base.

However, many commercial applications do not distinguish between the problem and solution part of a case because in many application domains there is no such a priori distinction. In this instance, a case just records a piece of experience, represented by a set of attributes. The distinction on problem and solution is determined during runtime of CBR application.

Cases can be distinguished as homogeneous, where they have the same representation (attributes), and heterogeneous, that is they have different attributes but may share some.

There is a lack of consensus within the CBR community as to exactly what information should be in a case. However, two pragmatic measures can be taken into account in deciding what should be represented in cases: the functionality and the ease of acquisition of the information represented in the case (Kolonder, 1993).

3.1.2 Storage of Cases

Case storage is an important aspect in designing efficient CBR systems in that, it should reflect the conceptual view of what is represented in the case and take into account the indices that characterise the case. The case-base should be organised into a manageable structure that supports efficient search and retrieval methods. A balance has to be found between storing methods that preserve the semantic richness of cases and their indices and methods that simplify the access and retrieval of relevant cases. These methods are usually referred to as case memory models. The two most influential case memory models are the dynamic memory model of Schank and Kolonder, and the category-exemplar model of Porter and Bareiss.

The Dynamic Memory Model

The case memory model in this method is comprised of memory organisation packets or MOPs. MOPs are a form of frame and are the basic unit in dynamic memory. They can be used to represent knowledge about classes of events using two kind of MOPs:

(1) Instances representing cases, events or objects,
(2) Abstractions representing generalised versions of instances or of other abstractions.

The case memory, in a dynamic memory model, is a hierarchical structure of episodic memory organisation packets (E-MOPs) (Kolonder, 1993), also referred to as generalised episode (GEs) (Koton, 1989) developed from Schank's more general MOP theory (Schank, 1982). The basic idea is to organise specific cases which share similar properties under a more general structure (i.e. a generalised episode). A GE contains three different types of objects: norms, cases and indices. Norms are features common to all cases indexed under a GE. Indices are features which discriminate between a GE's cases. An index may point to a more specific generalised episode or to a case, and is composed of an index name and an index value.

The case-memory is a discrimination network where nodes are either a GE, an index name, index value or a case. Index name–value pairs point from a GE to another GE or case. The primary role of a GE is as an indexing structure for storing, matching and retrieval of cases. During case storage when a feature (i.e. index name and index value) of a new case matches a feature of an existing case a new GE is created. The two cases are then discriminated by indexing them under different indices below the new GE (assuming the cases are not identical). Thus, the memory is dynamic in that similar parts of two cases are dynamically generalised into a new GE, the cases being indexed under the GE by their differences.

However, this process can lead to a explosive growth in the number of indices as case numbers increase. So for practical purposes most CBR systems using this method limit the number of permissible indices to a limited vocabulary.

The Category-Exemplar Model

This model organises cases based on the view that the real world should be defined extensionally with cases being referred to as exemplars (Porter and Bareiss, 1986). The case memory is a network structure of categories, semantic relations, cases and index pointers. Each case is associated with a category. Different case features are assigned different importance in describing a case's membership to a category. Three types of indices are provided, which may point to a case or a category:

1. Feature links that point from problem descriptors (features) to a case or category,
2. Case links that point from categories to its associated cases,
3. Difference links pointing from categories to the neighbouring cases that only differ in a small number of features.

A feature is described by a name–value pair. A category's exemplars are stored according to their degree of prototypicality to the category. Within this memory organisation, the categories are inter-linked within a semantic network containing the features and intermediate states referred to by other terms. This network represents a background of general domain knowledge that enables explanatory support to some CBR tasks. A new case is stored by searching for a matching case and by establishing the relevant feature indices. If a case is found with only minor differences to the new case, the new case may not be retained, or the two cases may be merged.

3.1.3 Retrieval of Cases

The retrieval of cases can be done informally, where the user browses and selects a relevant design case, or formally, where the system accepts a new problem definition as input and presents a set of relevant design cases s output. In last case the comparison of cases and determination of usefulness are performed.

Remembering experiences which are similar to a new problem situation is required in the problem solving process. The cases are compared by comparing their respective problem situations with the introduced new one to estimate the usefulness of a past problem according to the core CBR assumption that similar problem has a similar solutions. The cases are retrieved from the case base in accordance with their similarity to the new situation. Similarity is a key notion of CBR.

The usage of the term *similarity* in the area of case-based reasoning focuses on similarity as a fuzzy relation between two objects or their respective representations, the cases. Since it is intended to adapt available knowledge about old cases to solve problems in new ones, the similarity measures to be constructed depend on these intentions and both case representations. This leads to similarity measures for two cases.

Case-based reasoning will be ready for large scale problems only when retrieval algorithms are efficient at handling thousands of cases. Unlike database searches that target a specific value in a record, retrieval of cases from the case-base must be equipped with heuristics that perform partial matches, since in general there is no existing case that exactly matches the new case.

There are two different approaches to similarity determination in CBR.

The computational approach which is based on computing an explicit similarity function for all cases in the case base, and the representational approach using structured memory of cases. Some techniques attempt to combine these approaches.

In the computational approach the current problem is matched against the problems stored in the case base during the retrieval procedure. Retrieval from the case base is based on the vague matching of information entities of the newly introduced problem and problems from past cases. Matching is the process of comparing two cases to each other and determining their degree of similarity. Degree of similarity is assessed by a numeric computation and results in a single number which is intended to reflect all aspects of the similarity (Stanfill and Waltz, 1986; Aha, 1991, Voss, 1995).

For the representational approach, the case base is pre-structured. Retrieval is by traversing the index structure (Schank, 1982; Kolonder, 1984). Cases that are neighbours according to the index structure are assumed to be similar.

Some clarifications about notion of similarity have been presented. Burkhard (1998) introduced the notion of acceptance of cases. This is directly linked to the subjective notion of usability.

3.1.4 Reuse of Experience

Usually, new situation rarely match old ones exactly. It is necessary to adapt an old solution to fit a new situation. In design tasks, even a small difference between the current problem and the most similar case may require adaptation.

Adaptation compensates for the differences between an old situation and a new one. Thus, it tries to fit an old solution to a new situation.

There are three general kinds of adaptation:

1. Parametric adaptation that corresponds to the substitution, instantiation or adjustment of parameters.
2. Structural adaptation that revises a retrieved solution by applying adaptation operators or rules to solve a new problem.
3. Generative adaptation that reuses and adapts problem-solving episodes by replaying their derivation.

An ideal set of adaptation rules must be strong enough to generate complete solutions from scratch, and an efficient CBR system may need both structural adaptation rules to adapt poorly understood solutions and derivational mechanisms to adapt solutions of cases that are well understood.

Several techniques, ranging from simple to complex, have been used in CBR for adaptation. These include:

1. Null adaptation, a direct simple technique that applies whatever solution is retrieved to the current problem without adapting it. Null adaptation is useful for problems involving complex reasoning but with a simple solution. For example, when someone applies for a bank loan, after answering numerous questions the final answer is very simple: grant the loan, reject the loan, or refer the application.
2. Parameter adjustment, a structural adaptation technique that compares specified parameters of the retrieved and current case to modify the solution in an appropriate direction. This technique is used in Bain (1986), which recommends a shorter sentence for a criminal where the crime was less violent.
3. Abstraction and respecialisation, a general structural adaptation technique that is used in a basic way to achieve simple adaptations and in a complex way to generate novel, creative solutions. The planning system in Alterman (1988) uses this technique.
4. Critic-based adaptation, in which a critic looks for combinations of features that can cause a problem in a solution (Sycara, 1987).
5. Reinstantiation, is used to instantiate features of an old solution with new features.
6. Derivational replay, is the process of using the method of deriving an old solution or solution piece to derive a solution in the new situation. For example, in Mostow et al. (1989), the CBR application replays stored design plans to solve problems.
7. Model-guided repair, uses a causal model to guide adaptation, which is used for diagnosis and learning in auto mechanics, and (Goel et al., 1992) used in the design of physical devices.
8. Case-based substitution, uses cases to suggest solution adaptation as in Moorman and Ram (1992) a system for robot navigation.

3.1.5 CBR Applications Range

Case-Based Reasoning should be considered as a problem solving technique, whenever it is difficult to formulate domain rules and when cases are available. It should also be considered when rules can be formulated but require more input information than is typically available, because of incomplete problem specifications or because the knowledge needed is simply not available when solving the problem. Other indications to use CBR are: if general knowledge is not sufficient because of too many exceptions, or when new solutions can be derived from old solutions easier than from scratch. Many successful applications in these areas have proven the utility of this problem solving technique.

CBR systems have attracted a great attention in the legal and medical domains, especially as diagnostic and care systems, as well as in finance and insurance for customer support and credit assessment (Allen, 1994). In addition CBR has a diversity of applications in intelligent Web-based sales services (Wilke et al., 1998; Watson and Gardingen, 1999), in building and mechanical design (Mileman et al., 2000; Rivard and Fenves, 2000), in material science (Amen and Vomacka, 2001; Mejasson et al., 2001), in support to complex fault finding and troubleshooting (Aha et al., 1999) as well as in planning and real-time scheduling tasks (Bonzano et al., 1997; Coello et al., 1999).

Case-based reasoning applications can be classified by the type of task they perform. The main classification dimension distinguishes *analytic* and *synthetic* tasks.

Analytic problem solving is concerned with analyzing a given solution and deriving further inferences on these interpretations. In most situations, a problem is regarded as solved when an appropriate case has been found because the solution can be directly derived from that case. Examples of analytical problem solving are classification, case-based decision support, diagnosis, and information retrieval.

Classification applications are possible when the problem domain consists of two disjointed sets: a set of observation and a set of classes. A problem description is represented as a set of observations. The solution of a problem is selections of one or more classes. A classification system tries to determine to which class a new example case belongs to. Therefore, all the cases in the case base must be problem solution pairs where the solution contains the class. Troubleshooting is one typical application of classification. Examples of such applications are given in Heider et al. (1997) and Aha et al. (1999).

Case-based decision support (CBDS) helps in decision making by providing cases similar to an actual problem situation. The problem is a representation of a situation with missing information. The objective is to complete the description of that situation during problem solving to satisfy a certain demand for information (Lenz et al., 1998). Help-desk applications relying on documents are a prominent example of this application type.

Diagnosis applications are, in essence, of classification type. Diagnosis can be considered as a generalisation of classification in the sense that observations are not necessary known at the beginning but have to be inferred. Diagnosis and CBDS systems both deal with incomplete information. The difference, here, is that diagnosis not only deals with incomplete information, it also considers the costs of ascertaining further symptoms to further complete the information. Diagnosis systems are widespread in the medical domain (Koton, 1989; Heckerman, 1991; Schwartz et al., 1997) and in law (Ashley, 1990).

Another example of analytic tasks is *case-based information retrieval.* Information retrieval is somewhat similar to case-based decision support, but focuses on content-oriented document search. Its goal is to find useful documents to support problem solving, for example, searching the World Wide Web for products where some essential properties are not known initially, but the intended use of the product is. Examples are given in Watson and Gardingen (1999) and Wilke et al. (1998).

Synthetic problem solving tries to compose new pieces of knowledge that have not been available before, such as configuration and design, or planning.

Configuration is generally understood as the construction of an artefact from a given set of components respecting all the compatibility constrains based on knowledge of how the components can be connected. Typical CBR configuration systems are described in Hennessy and Hinkle (1991), Purvis and Pu (1995) and Rousu and Aarts (1996).

The goal of *planning* is to find a sequence of actions transforming a given initial situation into a desired goal situation. While the classical planning process consists mainly of a search through the space of possible sets of operators to solve a given problem, new problems are solved by reusing and combining plans or portions of old plans in case-based planning. Here, reusing already computed plans can be used to improve planning speed by adapting those plans. In a number of works (Kovacic et al., 1992; Munoz-Avila and Weberskirch, 1996; Bonzano et al., 1997), examples of case-based planning application are given.

Design introduces some degree of creativity because some components or even structural elements of the artefacts may not be present a priori. Design is once more subdivided, depending on the degree of creativity, into routine design, innovative design, and creative design. The use of experience is of particular importance in this area, but the reused experience almost never remains unmodified. The applications of CBR to design have ranged from largely informal domains. Case-based reasoning can support innovative design activity by reminding designers of previous experiences that can match new design situation, not necessary totally but only partially. Many applications of CBR for design are described in the literature, for example, systems for architecture design (Domeshek and Kolonder, 1992; Flemming et al., 1997; Voss, 1997), design of electro-mechanical devices (Narashiman et al., 1997), and software interface design (Tsatsoulis and Alexander, 1997).

Recently, CBR has been applied in chemical engineering for quality design (Suh et al., 1998), thermal analysis support (Nakayama and Tanaka, 1999), troubleshooting plant problems (Chaput, 1999), process control and plant supervision (Sanchez-Marre et al., 1997; Roda et al., 1999), ecological tasks (King et al., 1999), and supporting design in process engineering; more specifically equipment selection (Kraslawski et al., 1999a; Kraslawski et al., 1999b), and process synthesis and flowsheet design (Surma and Braunschweig, 1996; Pajula et al., 2001).

3.2 Models of CBR Process

In order to describe the CBR process, several general models have been proposed.

Kolonder (1993) considers CBR as a process containing the following steps: case retrieval as a primary step, proposing an area of solutions by extracting them from some retrieved cases. Next, adaptation - the process of fixing a past solution to fit a new situation, criticism of the received new solution, its evaluation based on external feedback, and finally storage of the verified solution of the current problem in the case base (Fig. 3.1).

Kolonder describes two main roles for CBR: to provide suggestions of a solution of a problem, and to provide context for assessing a situation (interpretive task). In problem solving, a ballpark solution to the new problem is adapted and then criticized. If the new solution fails, it is adapted again.

In an interpretive task, a ballpark interpretation is proposed, followed by a justification process that tries to create arguments for the proposed solution. The justification process compares and contrasts the situation with past cases, looking for similarities between the new situation and others that justify the desired result.

After reviewing many CBR systems Hunt (1995) proposed a basic structure for the CBR process, shown in Fig. 3.2. Once a case base has been created, the first step is to analyse the inputs in order to determine the features that are important for the selection of past cases in the case base. These features

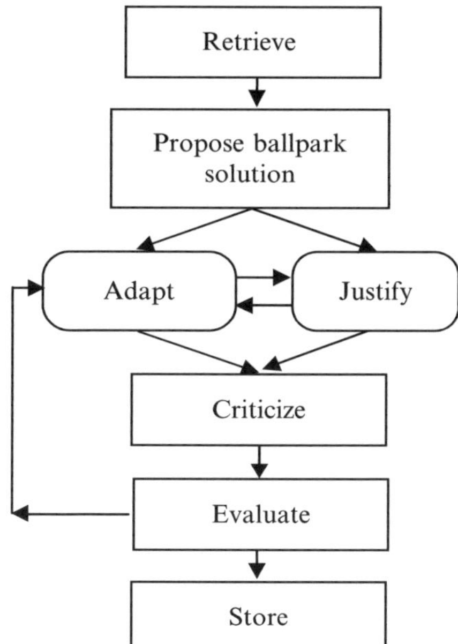

Fig. 3.1. CBR model according to Kolonder (1993)

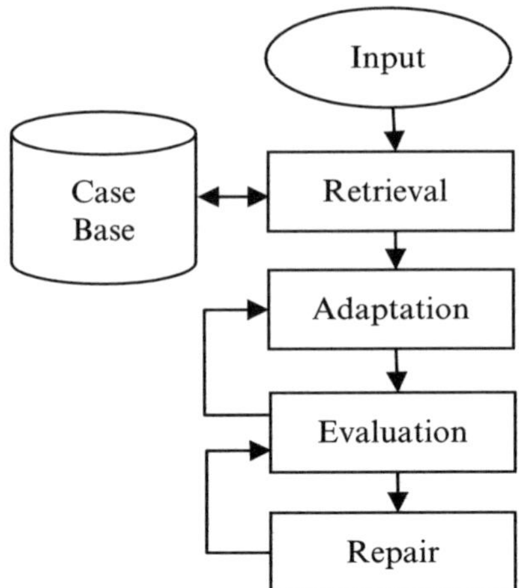

Fig. 3.2. Hunt's model of CBR (after Hunt, 1995)

are then passed to the retrieval step along with the initial inputs. The retrieval procedure uses this provided data to obtain a list of past cases which are similar to the current situation. Once the case has been retrieved, its solutions are modified during the adaptation step to fit the current problem. The obtained solutions must be evaluated to determine whether they provide a solution to the current problem. If the solution is accepted by the evaluation step, then it is presented as the solution to the problem and stored in the case base for future use. If some aspects of the current problem remain unsolved, then the solution must be repaired. Information about the reason of failure of the solution is used to guide the repair process.

Allen's model (Allen, 1994) includes five steps: presentation of the current problem, retrieval of the closest-matching cases stored in a case-base, its adaptation for generating a solution for the current problem, validation of the solution through feedback, and updating of the case base with the validated solution (Fig. 3.3).

Aamodt and Plaza (1994) introduced a model which consists of the following phases: retrieve the most similar cases, reuse the cases, revise the proposed solution, and retain the new solution as a part of a new case.

This model is commonly called the R^4 model of CBR, because the processes involved in this model can be represented by a scheme comprising the four REs, shown in Fig. 3.4. Each step involves a number of more specific steps, for example, retrieve includes identify, search, initially match and select (Aamodt and Plaza, 1994).

An initial description of a problem defines a new case. This new case is used to retrieve a case from the collection of previous cases. The retrieved case is reused to propose a solved case, i.e. a suggested solution to the initial problem. Through the revise phase this solution is tested for success, e.g. by

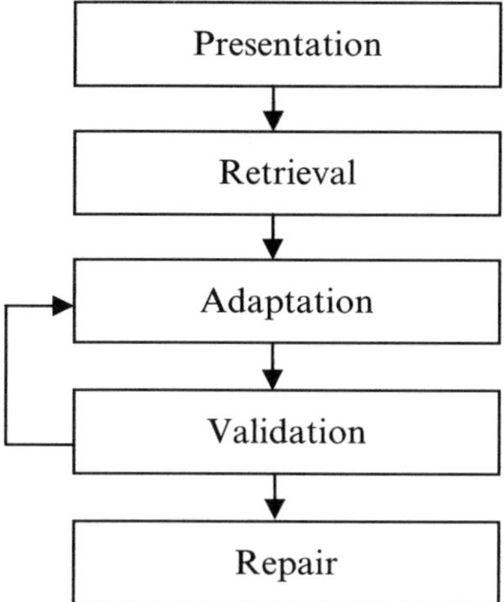

Fig. 3.3. Allen's model of CBR (after Allen, 1994)

being applied to the real world environment or evaluated by an expert, and repaired if failed. If the solution is accepted the tested/repaired case is retained for future use, and the case base is updated by a new learned case, or by modification of some existing cases.

General knowledge usually plays a part in CBR processes. General knowledge means general domain dependent knowledge, as opposed to specific knowledge recorded by cases.

However, these models assume that the case base is ready "at once" for case retrieval, and ignore the fact that case base building is also an important CBR task. Finnie and Sun in a recent work (Finnie and Sun, 2003) have considered the process of preparation of case bases. They extended the model of Aamodt and Plaza by adding a new step: repartition, which builds a satisfactory case base based on utilizing similarity relations to the possible world of problems and the world of solutions (Fig. 3.5). This step represents the process of case acquisition. The authors declared that the repartition step provides the theoretical foundation for case retrieval, because of the one-to-one correspondence between the partition and the similarity relations. Thus, case base building and case retrieval can be treated as similarity-based reasoning in a unified way (Sun et al., 2004).

In the given models it is assumed that case data are already structured according to some representation. However, in many real world problems, especially those of chemical engineering, the data format is not uniform and information is distributed among many sources located in different data bases.

An attempt to integrate the data analysis techniques into a CBR system has recently been made in Lau et al. (2003).

Liew and Gero (2002) have extended the basic idea of case-based reasoning and introduced a model of situated case-based reasoning (situated CBR)

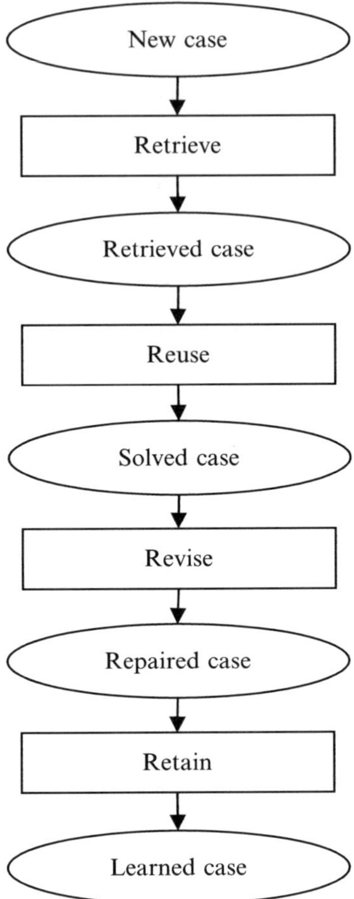

Fig. 3.4. The model of CBR process (modified Aamodt and Plaza, 1994)

based on a model of constructive memory (Fig. 3.6). In the situated CBR model, instead of focusing on just the design problem and finding a solution, emphasis is given to the environment within which the problem is framed. The model interprets the environment according to the current situation and the problem is framed accordingly. This interpretation is dependent on the current environment, the internal state of the situated CBR system and the interactions between the system and the environment. The internal state of a situated CBR system is defined by its content. This content is made up of individual entities that are classified either as experience or knowledge. Interactions between the system and the environment define different interpretations of the environment according to different interpretations of the selected entities used for memory construction.

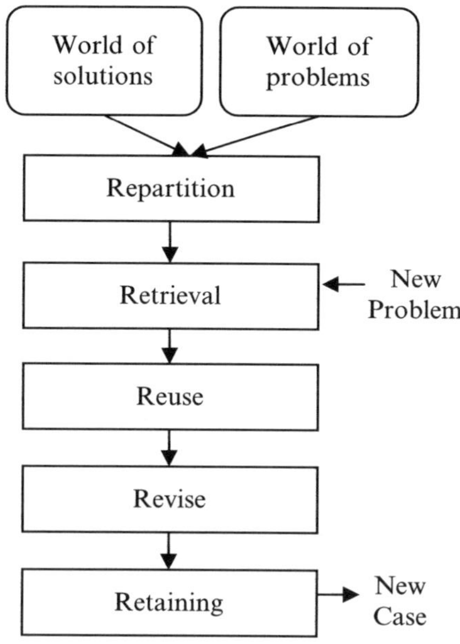

Fig. 3.5. CBR process model according to Finnie and Sun (2003)

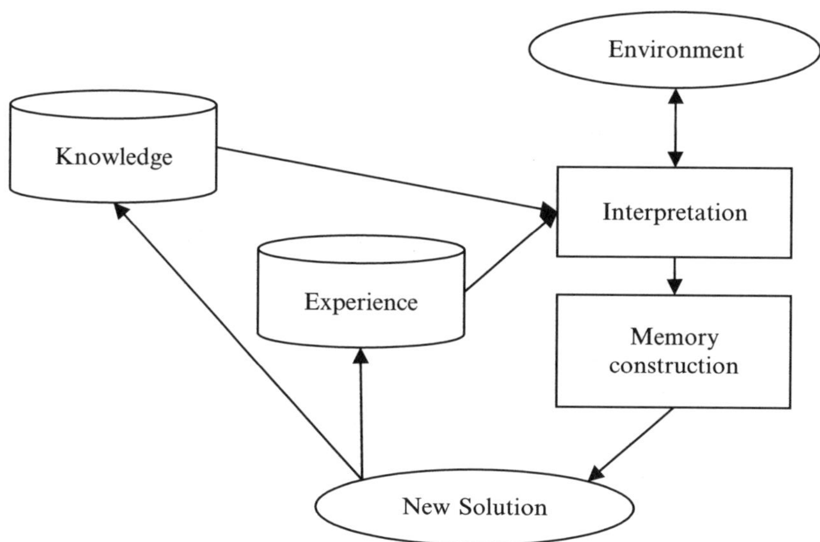

Fig. 3.6. Situated CBR model (after Liew and Gero, 2002)

The memory of an experience and/or knowledge (entities) is constructed according to an interpretation of the environment and an interpretation of the selected entities relevant to the problem at hand. Rather than adapt a selected case to new design specifications, the selected entities are interpreted according to the interactions between the system and the environment. These interactions provide a specific view (interpretation) of the relationship between the design specifications and the environment. This view dictates another interpretation of the environment that can introduce new specifications (Liew and Gero, 2002).

In summary, it can be said that most existing models consider a case structure as a solid one, remaining unchanged during the overall reasoning process. The model of Finnie and Sun, which includes case acquisition, nevertheless relies on defined similarity relations in available sets of problems and solutions. If the environment has been changed, these relations do not remain unchanged. The gathered experience must be interpreted in accordance to a specific current design task. Liew and Gero proposed such interactions between environment and experience, but their approach is based on a memory construction model and using specific knowledge that is far distant from the conventional CBR paradigm.

A novel model of the CBR process, which can take into account all aspect of CBR activities, would be useful extension of the conventional CBR paradigm. Such model could support the design process at different levels of abstractions and in changing design tasks based on different interpretation of a gathered set of experience.

3.3 Case-Based Design Support Methodology

The extended model of the CBR process is suggested to compose six steps: collect, constitute, compile, compare, correct and check (Fig. 3.7).

The first step is the collection of necessary data from the environment. Under environment is understood a set of information sources that is not part of the CBR system. Depending on design task, the appropriate specific data are extracted from the sources. The next step is to constitute a case structure that is best suited to describe the area of the specific design problem to be solved. Once the case representation has been obtained the collection of relevant data is processed to create the case base of the specific case structure. During the next step, a new problem to be solved is introduced according to established case representation and compared with past cases from the created case base. Once the most similar case has been retrieved, its solution is corrected in the following step. The corrected solution is the subject of validation checking. The checked cases can be stored in the environment to extend its scope.

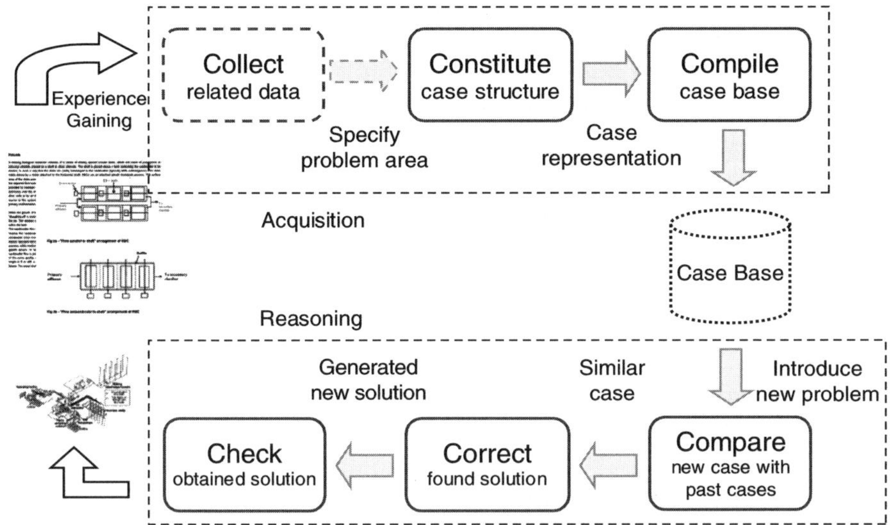

Fig. 3.7. The six-steps model of CBR process

Overall the CBR process is decomposed into two groups of activities: Acquisition of existing design information in order to compile the case-base, and reasoning, finding a solution for a specific design problem.

The actions of case-based design assistance are:

Acquisition

1. Represent complex design cases
2. Augment cases with generalized design knowledge
3. Formalize a typically informal body of knowledge
4. Transform from formalized design representation into memory organization

In situations where only an informal body of knowledge is available, the design assistance may focus on tasks that can be formalized.

Reasoning

1. Introduction of a current problem represented according to case formalization
2. Similarity determination
3. Ranking the cases and retrieval of a set of most similar cases
4. Solution proposal to give alternative candidate solutions among the selected and ranked cases
5. Solution modification to make changes in the design proposal and evaluation to verify the feasibility of the proposed solution and its satisfaction of numerical and logical constraints

6. Incorporation of the generated solution for the current problem to the case base to improve the capability of the case base

The knowledge structures involved in case-based design assistance are:

- Case representation, which is used to create cases from information sources;
- Similarity measures, which are used to compare cases with an input problem;
- Adaptation rules, which are required to correct a found solution to fit the current problem.

The lack of formal knowledge in design affects both in ability to define a formal and consistent representation of design case and the role of adaptation as a human-cantered activity or an automated process.

3.3.1 Collection of Relevant Data

First, the data containing experience which could be used to solve a design problem of a certain type are collected. Experience can be stored in different information sources.

Information sources are, for instance, domain experts, handbooks, manufacturer newsletters, specialized literature, and magazines. Of course, not every information source is relevant to the system. The relevant information sources form the borders of the environment of the system. Simple structured information sources make it a lot easier to maintain a particular system. Thus, to ease extraction a requirement on the information sources is that they are well structured. If the data are not structured there is a need to apply methods of data mining to recognize relevant information.

Aamodt and Nygard (1995) describe a model that clarifies the distinction between *data, information*, and *knowledge.*

According to their model, data are syntactic entities, patterns with no meaning. They are input to an interpretation process, which is the initial step of decision making. Information is interpreted data, data with meaning. It is the output of data interpretation as well as the input to, and output from, the knowledge-based process of decision making. Knowledge is learned information, it is the output of a learning process made ready for active use within a decision process.

The essential role of knowledge, in general, is therefore to play the active part in the processes of transforming data into information (referred to as *data interpretation*), deriving other information (referred to as *elaboration*), and acquiring new knowledge (referred to as *learning*).

Information must be transformed into knowledge to be accessible by the case-based design assistance. This knowledge is used to constitute the case representation.

3.3.2 Representation of Complex Design Cases

Design usually involves the development and understanding of complex systems. The complex representation needed to adequately capture a design case has introduced challenges to CBR systems. A design case often is supplemented with many related knowledge which also have to be represented.

Process design case representation is an abstraction of experience into symbolic and numeric form easily used by a computer assistant to effectively retrieve and evolutionarily modify previous models to meet a new design requirement.

The problems associated with appropriate representations and suitable information processing chains for engineering design consist of the following main issues:

- Topologically and parametrically different design solutions, for which envelopes or alternative enumeration schemes are not known
- Verbal, symbolic and numerical pieces of information in different design models and at various degrees of detail.

Thus, the design information can be defined as structural and parametric.

Representation of the design data requires various models because design content involves topological, geometric, and physical properties and their relationships. In additional to the representation of complex design cases, there is a need in domain knowledge that is represented separately – as rules, models or constrains. Therefore, design knowledge may include causal models, state interactions, heuristic models, heuristic rules and geometric constrains. These are generalized design knowledge.

Complex cases can be characterized as cases which (Gebhardt et al., 1997):

- May have to be cut out of large data models
- May not be described sufficiently in terms of attributes but have to be represented structurally
- Contain variables that do not describe a problem or a solution; the role of variables can be dynamically changed
- May be useful in multiple ways and allow for more that one representation.

Multiple case interpretations require a flexible combination of several similarity functions. Different aspects of a case (features, the structure of cases) may have to be jointly considered for retrieval and adaptation.

The proposed way to build a case base which can represent diverse design data is the consideration of information entities. A case is set of information entities. The number of information entities in a case may be variable.

The representation of an entity is based on the assumption that any design entity can be represented by a finite set of features and their relationships.

An entity description includes the list of features F, the set of relations between them R, and the set of feature values V. The representation may be extended by including numerical attributes of the features of an entity, W.

The attributes usually reflect a degree of importance of the corresponding feature in the description.

Features may be of various types. They can be expressed by numeric, vector, logical or symbolic values, as well as sets and graphs. The main advantage is that a feature may be represented by a new information entity. Thus, a case may contain a hierarchy of entities.

A new problem to be solved is also represented in the form of information entities. The appropriate list of features, as well as the set of their relationships is defined according to a frame of the problem of a design task.

Taken together, complex case representations cause increased computational expense in retrieval and adaptation. To provide a proper answer for real world application, efficient memory organization directly regarded to the applied reasoning algorithm is essential.

3.3.3 Memory Organization

Conceptually, any real or abstract entity (such as a fixed bed reactor, distillation column, heat exchanger or its mathematical representation) is considered as an object that can be referenced by a unique identifier. Objects have one or more attributes to express an entity's properties.

All objects which share the same set of features and relationships can be considered as an instance of a class.

The values of the features of an object can be pointers to objects themselves to enable the construction of semantic links, connections between it and the object at the lower level. Any subclass can be derived from another class by inheriting all features and relations. This inheritance relationship allows the construction of process hierarchies.

The object-oriented representation scheme allows a hybrid of familiar storage techniques: tables, trees, relations. One common technique for process design memory organization is table storage. It is simple to use in applications, but it does not allow the multi-level principle of process design case to be followed. To overcome the difficulty each instance of a class corresponding to a specific level of abstraction in the design representation is represented by the tables of a certain structure. The tables are linked with the upper level of abstraction by means of connection tables.

3.3.4 Compilation of Case Base

The information describing the design cases is collected and next the corresponding data are stored in free format in several data sources. The design data is processed and the representation structure of the cases is built. The structure of the cases is constructed according to the requirements of the design task.

All information stored in the collection phase is re-organized to make it suit the constructed case structure. The data is transformed to a set of cases.

The cases are represented by a uniform format. It can be XML-based (Extended Mark-up Language) representation, relational tables format, or others.

3.3.5 Comparing Cases

When the case base has been created, the CBR system is used to find a solution to the actual problem introduced to the system.

The quality of the proposed solution for the design problem mainly depends on the similarity measure that is used to retrieve the similar cases thanks to the fact that the retrieved cases constitute the starting point for finding the solution of the actual problem.

The computational approach for the similarity measure, which implies computing an explicit similarity function for all cases in case base, is more general and it is utilized in the developed methodology for problem solving support of the design tasks.

Cases may contain entities which have no counterpart in the new problem. It is also possible that some entities of a new problem are not present in the case base. Moreover, entities may include various numbers of features and the features often have different values. Three levels for similarity measurement can be distinguished: cases, entities and features. But both cases and features can be represented as entities. A case is an entity containing one feature represented as a set. Hence, the similarity measure is uniform for each level.

In order to cope with design cases that have different structure representations in the case base the general similarity concept has been developed. The general similarity concept is described in the following chapter.

3.3.6 Correction of Found Solution

In design tasks, even small difference between the current problem and the most similar case may require correction of the found solution – adaptation. Parametric adaptation, where design parameters of the retrieved solution are adjusted to remove the gap between the past design and the current problem, are considered in the work.

An adaptation procedure is based on the assumption that not only the most similar case can be used but a set of cases located nearby the current problem in the problem space can also be used. The key assumption that a similar problem has the similar solution means that solutions of similar problems are located nearby each other. The distances between the new solution and solutions of the most similar problems must correspond to the distances between the current problem and similar problems under consideration.

Since the solutions are also described by a set of information entities, the computation of distance is also based on the general similarity concept.

3.3.7 Checking Obtained Solution

The solution proposed by the correction phase is the subject for evaluation of whether it provides a proper solution to the current problem. This phase requires external knowledge (expert assessment) for evaluation. If suitable evaluation criteria and problem models are available then the case-based design assistance is able to perform this task without user interventions. But often interaction between the assistance and the designer is required. Simulation and parameter calculation might need to be carried out.

If the solution is approved, it is stored in the environment in some way to extend the scope of information sources. This solution together with the current problem can be use in future problem solving as a new piece of experience.

3.4 Summary

A model of the CBR process in the framework of the conventional CBR paradigm which can support the design process at different levels of abstractions and in changing design tasks based on different interpretation of gathered set of experience is proposed. It includes a cycle of 6-Cs – collect, constitute, compile, compare, correct and check. The model implies not only runtime reasoning but also runtime case acquisition that differs from most other CBR models. It is necessary in a changing design environment when the data structure is being changed during evolution from less to more abstract levels of the design process. The design data can be represented in various forms – vectors for composition of chemical compounds, graphs for distillation systems, sets for equipment specifications. In order to cope with different data structures and various information sources, a general approach for representation of the design artefact, which can be used in any stage of design process, and for comparison of these representations, are required.

4

Similarity and Adaptation Concepts

The quality of a solution to a problem in case-based design support is highly dependent on the representation of the design artefact and the comparison of design situations based on their similarity. The design case is represented as comparable with past experiences. It is also stated that the design of any complex chemical process or a new chemical product can be considered at various abstraction levels. In order to support the design activity, at any particular stage the representation of the design situation or artefact must be task independent. A comparison of different design cases requires a clear statement of similarity. Similarity assessment has to proceed over complex representation of design case. The design data may include many data models that require a flexible combination of similarity measurements. In this work a general concept of similarity that can cope with various formats of design data has been developed. Uniform representation of the data of the solution and problem parts of the cases and applying a similarity concept for solutions leads to an original idea of adaptation which is task independent as well. This chapter describes both the similarity and adaptation concepts.

4.1 Retrieval Method and Similarity Measures

In CBR systems, the quality of the results mainly depends on the similarity measure that is used to retrieve the similar cases. It is so as the retrieved cases constitute the starting point for finding the solution of the actual problem. During the retrieval procedure, the current problem is matched against the problems stored in the case base. Matching is the process of comparing two cases to each other and determining their degree of similarity. If the case is represented as a set of features and their values, the similarity measure

Y. Avramenko and A. Kraslawski: *Case-Based Design*, Studies in Computational Intelligence (SCI) **87**, 71–97 (2008)
www.springerlink.com

between two cases can be determined by the following operations (Kolonder, 1993):

(1) Find the corresponding features in the compared cases.
(2) For every feature, compute the degree of similarity between the corresponding features of the cases.
(3) Multiply the obtained values by the corresponding coefficient reflecting the importance of the feature (weight of importance) and sum them to get overall similarity value.

There are several approaches to compute the degree of similarity of the features based on distance on a quantitative scale, a position in a hierarchical structure, comparison of qualitative definitions (this is determination of similarity between values expressed on qualitative scale – like 'high', 'low', etc.) and structural comparison.

4.1.1 Quantitative Distance

Most of engineering parameters is of quantitative nature. The value of the parameters can be expressed as a number, a vector, a set or an ordered sequence. The computation of similarity between numeric values is based on nearest-neighbour algorithm. The nearest-neighbour method relies on a metric. A metric is merely a function giving a scalar distance between two arguments.

Numeric Vectors

Distance is one measure of vector similarity. The larger the distance between two vectors, the smaller their degree of similarity.

The most common metric is the Euclidean distance formula that for vectors a and b in n-dimensional space is:

$$d(\mathbf{a}, \mathbf{b}) = \sqrt{\sum_{i=1}^{n} (a_i - b_i)^2} \qquad (4.1)$$

More general metric for n-dimensional vectors is the Minkowski metric that is also referred as the L_k norm (Schalkoff, 1992):

$$L_k(\mathbf{a}, \mathbf{b}) = \left(\sum_{i=1}^{n} |a_i - b_i|^k \right)^{1/k} \qquad (4.2)$$

When $k = 2$ the norm (4.2) transforms to the Euclidean formula. The norm with $k = 1$ is called the city block (or Manhattan) distance, the shortest path between **a** and **b**, each segment of which is parallel to the coordinate axis.

The block city and the Euclidean formulas are wide-spreaded as the similarity measures in CBR applications. The block city is used for numeric values, for example, in Amen and Vomacka (2001) and Pajula et al. (2001), or with modifications of the criteria in Roda et al. (1999) and Sanchez-Marre et al. (1997). The Euclidean metric is employed as a similarity function for numeric features, for example, in Avramenko et al. (2002).

Actually in some applications where the case is simply described as a list of numeric features, the cases are represented as vectors of features and overall similarity between the cases is calculated in similar way, utilizing the block city or the Euclidean metrics, or even high order Minkowski norms (Althoff et al., 1995).

It should be noticed that the components of vectors must be normalized over all set of data in order to get proper value of similarity measure.

Sets

The similarity measures of the set rely on a metric function as well.

The distance between sets A and B is defined usually by Tanimoto metric (Duda et al., 1998):

$$d(A, B) = 1 - \frac{|A \cap B|}{|A| + |B| - |A \cap B|} \tag{4.3}$$

The metric is mostly used for solving problems where the elements of the set are equally important, and there is no natural notion of graded similarity. This metric (with additional coefficients, specific for the task studied) is used in reference (Surma and Braunschweig, 1996) to compute the aggregation similarity between the components of flowsheet, and to determine the connection similarity that focuses on the stream connections between its components.

In Nakayama and Tanaka (1999) the authors used the sets distance measure to determine the similarity between the cases for the feature 'phase' that is represented as a list of terms such as glass, gum or crystal.

Another related measure is the Levenshtien distance (Schalkoff, 1992):

$$d(A, B) = \max(|A|, |B|) - |A \cap B| \tag{4.4}$$

The formula was applied in reference (Avramenko et al., 2002) to determine the similarity between the features expressed as textual variables, which is represented as a list of keywords. The most significant domain notions were collected in one set, specific for each textual feature type, and hence each context of feature is a subset of those full set of keywords.

Textual Strings and Sequences

The string in general can be represented as a set of words or symbols. Very often the ordering of the elements is important, and the string must be considered as a sequence of symbols.

For such sequences, element-by-element comparison is performed and similarity measure is based on computation of a Hamming distance. The Hamming distance of two sequences of equal length is calculated by counting the character positions in which they differ.

Another similarity metric relies on determination of so-called edit distance between two strings. The edit distance between strings A and B determines how many fundamental operations are required to transform A into B (Duda et al., 1998). These fundamental operations are:

(1) Substitutions: a symbol in A is replaced by the corresponding symbol from B.
(2) Insertions: a symbol in B is inserted into A, thereby the length of A is increased by 1.
(3) Deletions: a symbol in A is deleted, thereby the length of A is decreased by 1.

The value of the edit distance is calculated as the minimum number of basic operations needed to transform A into B.

Sometimes the feature has a unique value and no partial match is allowed. That is impossible to determine the grade of similarity for such values. In this case, only an exact match is considered and the value of the similarity measure is either 1, if one feature is completely matched to another or 0 otherwise.

$$sim(a,b) = \begin{cases} 1, & a = b \\ 0, & a \neq b \end{cases} \tag{4.5}$$

The exact matching is quite spread in many CBR applications. In Amen and Vomacka (2001) it has been applied for any symbolic parameter of the case. In Nakayama and Tanaka (1999) it has been used for the determination of similarity of sample's form and parameters of production process, and in Avramenko et al. (2002) – for comparing the class of reaction rate, type of catalyst and code of catalyst composition.

Hierarchical Tree

In some cases, the all possible values of a feature can be grouped in some classes and a hierarchical structure can be built to show the relations between the classes. Each value is represented by a dangling node in the hierarchical tree. The similarity measure between two values is based on the level of the hierarchical tree where the nearest common node can be found. The more close the common node, the better the match. In order to calculate the similarity in numeric form the corresponding values to all nodes in the tree are assigned. A score of 0 means the least similar, and 1 is for the most similar cases.

However, when the features can be classified in the several different ways or some values may be related to different classes simultaneously, the use of hierarchy could be ambiguous. In consequence it could result in the assignment of the different values of the similarity when comparing the same two features.

The determination of similarity based on hierarchy has been used in several applications in chemical engineering. In Nakayama and Tanaka (1999) the polymeric materials are classified hierarchically from the viewpoint of chemical structures. Each internal node of a classification tree corresponds to the category of a material. There is also used hierarchical definition of similarity for sample information and for purpose of measurement in this work. The hierarchical structure is often used for similarity determination of substances based on their chemical structures (Pajula et al., 2001; Avramenko et al., 2002). The comparison based on hierarchy can be also used to define similarity of processes and equipment units, e.g. separation process and separation equipment (Pajula et al., 2001). The closer the separations are to each other, in the tree structure, the more similar they are. In Mejasson et al. (2001), the classification tree approach was used to classify the component type. At the root of the tree, the components are divided into three groups: mechanical, electrical and 'soft'. Each node has a value that determines the similarity of two components.

4.1.2 Qualitative Comparison

Another way of measuring degree of similarity is determination of the distance between two values on a qualitative scale (Kolonder, 1993). The distance between the values belonging to the same qualitative category is considered equal to 0. Otherwise, the distance between two qualitative values is determined by the number of other categories separated the values from each other on the qualitative scale. The more qualitative categories separate two qualitative values, the lower the similarity measure. An integer value can be assigned to each qualitative category and the similarity measure between two qualitative variables can be computed by determination of distance between these integer numbers.

The use of qualitative comparison for the calculation of the degree of similarity is advantageous when small differences in features values are irrelevant to the degree of match. However, a problem may arise when similarity measure is inaccurate at the edges of the ranges.

The qualitative scales are widely used in many CBR applications. In Mejasson et al. (2001) the qualitative scale is used to represent the operating parameters, such as temperature etc. The authors have used five-degree scale: very low, low, medium, high and very high. The maximum distance of 1 is between the highest and the lowest categories (Table 4.1).

There are two qualitative factors describing the value and reliability of the design case in application (Pajula et al., 2001). These are technical maturity and performance (goodness) factor. The latter describes the proven efficiency of the design and has six gradation levels (it is numbered from 0 to 5 in order to compute a similarity measure).

Table 4.1. Example of distance measure for qualitative values (Mejasson et al., 2001)

	Distance = 1	Distance = 0.25	Distance = 0.5
Very Large	▲		
Large	│		▲
Medium	│		│
Small	│	▲	●
Very Small	●	●	

Qualitative variables are used also in the reference (Chaput, 1999) to describe the troubleshooting plant situations.

4.2 General Similarity Concept

A comparison of different design cases requires a clear statement of similarity. Similarity assessment has to proceed over complex representation of design case. The design data may include many data models that require a flexible combination of similarity measurements. Introduced general concept of similarity can cope with various formats of design data. Uniform representation of the data of the solution and problem parts of the cases and applying a similarity concept for solutions leads to an original idea of adaptation which is task independent as well.

4.2.1 Basic Notions

In the framework of the concept, an entity from the real world (that can be a substance, a phenomena or a process) is considered only in the form of its description – information content that characterizes an entity by a finite set of its properties and their relations.

Definition 1. *An attribute is any structural property (e.g. physical, mechanical, geometrical, or chemical) that can be observed or measured by specific means (e.g. through use of an instrument).*

Definition 2. *A function property is the behaviour that an artifact displays when it is subjected to a situation. The collection of all functions observed in different situations is the functional description of the artefact.*

Definition 3. *An entity E is a three-tuple and is defined as follows:*

$$E = \langle F, V, R \rangle \tag{4.6}$$

where F *–a finite set of features of an entity reflecting their nature;*
V *–a set of values of features;*
R *–a set of relations between features.*

A feature can represent both a attribute and a functional property of an artifact. A feature can be elementary and complex in nature. An elementary feature represents a property that cannot be defined in terms of other features. Complex feature is represented as an entity.

Definition 4. *The set of features F and the set of relations R form the structure S of entity E*

$$S = <F, R> \tag{4.7}$$

The representation of the structure may be extended by including numerical and perhaps symbolic attributes of the features of an entity. The attributes may reflect a degree of importance of the corresponding feature in the description. In such a case the structure is called weighted and is defined as follows

$$S = <F, R, W> \tag{4.8}$$

where W is a set of weights corresponding to the features of an entity.

Thus, the structure of an entity is a graph, each node of which may contain a characteristic index (e.g. weights of importance). Many entities might have one structure. Thereby a structure represents a *class* of entities, whereas a single entity is an *instance* of that class.

Definition 5. *If for two entities* $E_1 = <F_1, V_1, R_1>$ *and* $E_2 = <F_2, V_2, R_2>$, $F_1 \subseteq F_2$ *and* $R_1 \subseteq R_2$, *or* $F_2 \subseteq F_1$ *and* $R_2 \subseteq R_1$ *then the entities are denoted as structurally similar, otherwise the entities are structurally dissimilar.*

Further only the structurally similar entities are considered. Two entities E_1 and E_2 will be described as $E_1 = <S_1, V_1>$ and $E_2 = <S_2, V_2>$, where $S_1 = <F_1, R_1>$ and $S_2 = <F_2, R_2>$.

Definition 6. *Two features of different entities are denoted as* corresponding *if they are in the same relationship with other features in the structures. If the entities have different structures only relations belonging to intersections of sets of relations are taken into account.*

Definition 7. *Two entities are* similar *if all or some part of their corresponding features has identical values. Otherwise, when no part of the values match, the entities are dissimilar.*

If all corresponding values of entities E_1 and E_2 belonging to one class are matched, i.e. $S_1 \equiv S_2$ and $V_1 \equiv V_2$, the entities are called identical. Obviously, all identical entities are similar as well. When only a part of the corresponding values are different, the entities are partly similar.

Often, an exact match of values is not necessary, or it is not possible to assign an exact value for a feature, especially in design. It is subject of tolerance (acceptable deviation) for parameters, or acceptable ranges of values, or certain types of values.

Definition 8. *If all or a part of corresponding features of two entities have values that belong to certain classes (specific for each pair of features) of a specific classification, then the entities are called* conditionally similar. *The condition implies a set of rules that divide the features values into the classes.*

Explanations. The data of objects might be represented as numbers, symbols and schemes. Corresponding elements of the objects might belong to a numeric class (e.g. n-dimensional vector) or to other classes but necessarily for corresponding elements these classes should be common. Then the objects are regarded as similar under a certain classification. If one element is a number but its corresponding element is a text then the elements are not similar under the condition. Another example is range similarity. Numeric parameters of an object might simply be divided into micro and macro scales. If the values of different objects are all of micro scale then the object is similar under the condition, even if the corresponding values are not identical.

The notion of the similarity under the condition is very important and allows determining the similarity of entities even under uncertainty. With increasing specificity of the classification law, the similarity under condition approaches simple similarity. Similar entities are also similar under condition when each element creates own its specific class.

The notion of similarity is not useful enough in practice, as the entities will be similar when both all elements and only one element of the entities are identical. Therefore it is necessary to define the magnitude of similarity between two entities.

Definition 9. Degree of similarity *is the value expressed showing how much one entity is similar to another entity and is defined by a ratio of the number of matched features of two entities to overall number of features in the structure.*

4.2.2 Overall and Particular Similarity

For two entities A and B of one class (i.e. $\mathrm{S_A} \equiv \mathrm{S_B}$), defined by sets of values $V_A = \{a_1, a_2, \ldots, a_\mathrm{n}\}$ and $V_B = \{b_1, b_2, \ldots, b_n\}$, where equal indices determine the corresponding features of the entities, the degree of similarity (denoted as *sim*) according to definition 9 is defined as

$$sim(A,B) = \frac{1}{n} \cdot \sum_{i=1}^{n} ident_i, \qquad \text{where } ident_i = \begin{cases} 1, & a_i = b_i \\ 0, & a_i \neq b_i \end{cases} \tag{4.9}$$

When the entities have a weighted structure (4.8), each feature in the structure has a certain number, weight. The similarity degree of the entities with weighted structures is defined taking into account the values of weights w_i:

$$sim(A,B) = \frac{\sum_{i}^{n} ident_i \cdot w_i}{\sum_{i}^{n} w_i} \tag{4.10}$$

For conditionally similar entities analogically there is the notion of conditional degree of similarity, where the *ident* function is replaced by a class membership function:

$$\mathrm{sim}(\mathrm{A},\mathrm{B}|\mathrm{C}) = \frac{\sum_{i}^{n} \mu_i \cdot w_i}{\sum_{i}^{n} w_i} \tag{4.11}$$

$$\text{where } \mu_i = \begin{cases} 1, & a_i, b_i \in C_k \\ 0, & a_i \in C_k, b_i \notin C_k \end{cases} \quad (C_k - \text{some class})$$

In the last equation, C means the condition – a classification law; μ determines the membership of values a and b of corresponding features to a class. If the bounds of a class cannot be certainly determined, i.e. they are fuzzy, then the function $\mu(a_i, b_i)$ takes the values within the interval (0; 1). Further, increasing specificity of the classes to the situation when the values of the features of entity B build classes with only one certain member (corresponding values of entity B) but with fuzzy bounds, membership function μ get a new meaning. If b_i exactly belongs to a class (builds a class), then $\mu(a_i, b_i)$ shows how a_i is similar to b_i. This leads to the notion of degree of similarity for values of features.

When it is possible to say how similar two objects are, it is also possible to determine how different they are. Similarity and difference are opposite notions that characterize the same property but from different 'sides'. Therefore, degree of similarity l and degree of difference (distance) d supplement each other to 1:

$$l(a,b) + d(a,b) = 1 \tag{4.12}$$

Hence, the similarity l for two features' values a and b is defined as:

$$l(a,b) = 1 - d(a,b) \tag{4.13}$$

In contrast to the degree of similarity between entities, which is called overall similarity or global similarity, the degree of similarity between values of features is called particular similarity, or local similarity.

Degree of difference $d(a,b)$ (further called simply 'difference') shows how different two elements a and b are. The value of difference lies in the interval from 0 to 1 as well, where 1 corresponds to completely different elements. The equation for determination of difference depends on the data type of the values. The values of features might be represented as numbers, vectors, symbols, sets, graphics, and so on.

4.2.3 Difference Measurements

Let us introduce some symbols. The values of an entity's features, between which a difference value is determined, is denoted as a and b. The absolute difference of the features' values is denoted as Δ; the relative value of difference, that is the ratio of absolute difference to maximum difference, is denoted as d. The relative difference takes values within the interval (0; 1), and thus it is the target value.

Numeric Values

If the values of features belong to the class of real or integer numbers, then the difference, obviously, is determined by the absolute value of difference between the numbers:

$$\Delta = |a - b|, \quad d = \frac{|a - b|}{range} \tag{4.14}$$

where *range* – a range of values of variable a and b.

The range of values is defined over the set of all possible values of variables a and b. The set is assigned by knowledge domain or gathered base of objects of one type.

For n-dimensional vectors $\boldsymbol{a} = (a_1, a_2, \ldots, a_n)$ and $\boldsymbol{b} = (b_1, b_2, \ldots, b_n)$, the difference value is calculated using Euclidean metric:

$$\Delta = \left|\vec{a} - \vec{b}\right| = \left|\vec{d}\right|, \ \vec{d} = \sqrt{\sum_{i=1}^{n} (a_i - b_i)^2} \tag{4.15}$$

The absolute value of $\vec{d}$ is the distance between two points, defined by radius-vectors $\vec{a}$ and $\vec{b}$ in the space of these vectors.

If all coordinates of the vectors are equivalent, it means the coordinates are equally important and none can be emphasized, so it is necessary to carry out normalization: the real values of the coordinates are converted to relative ones, belonging to the interval (0; 1). Then the length of vector $\vec{d}$ is determined by the difference of the points' coordinates but not by the real great value of one or several coordinates.

The distance vector is determined in relative coordinates in n-dimensional space as follows:

$$\begin{aligned} \vec{d} &= (d_1, d_2, \ldots, d_n) = \left(\frac{a_1 - b_1}{range_1}, \frac{a_2 - b_2}{range_2}, \ldots, \frac{a_n - b_n}{range_n}\right) \in R^n \\ \left|\vec{d}\right| &= \sqrt{d_1^2 + d_2^2 + \ldots + d_n^2} \end{aligned} \tag{4.16}$$

The values $range_i$ determine the range of coordinates change that are assigned by either the knowledge domain or the gathered set of entities as it is for single numbers.

Explanations. There are two vectors $\vec{a} = (0;\ 50;\ 2,5)$ and $\vec{b} = (1;\ 100;\ 3,6)$. The absolute value of the distance vector of these vectors is determined mostly by second coordinates, as

$$\left|\vec{d}\right| = \sqrt{(0-1)^2 + (50-100)^2 + (2,5-3,6)^2}$$
$$= \sqrt{1 + 2500 + 1,21} \approx \sqrt{2500} = 50$$

So, the differences of the first and third coordinates are almost neglected when the difference value is calculated. But in the case of relative coordinates, where 1 corresponds to the maximum possible value and 0 corresponds to the minimum possible value (it might even be a negative value) of the coordinate the difference of all coordinates is taken into account. If in relative coordinates the vectors are $\vec{a} = (0;\ 0,4;\ 0,2)$ and $\vec{b} = (1;\ 1;\ 0,4)$, then

$$\left|\vec{d}\right| = \sqrt{(0-1)^2 + (0,4-1)^2 + (0,2-0,4)^2} = \sqrt{1+0,36+0,04} \approx 1,2$$

The ranges of coordinate values create an area in space R^n which, when converted to relative coordinates, forms n-dimensions unit cube. The cube contains all the vector-elements from the gathered set. Let us transform this area into a new space with basis vectors $e_1 = (1;0;\ldots;0)$, $e_2 = (0;1;\ldots;0), \ldots,$ $e_n = (0;0;\ldots;1)$. Each basis vector corresponds to the maximum change of the corresponding coordinate for the vectors of the gathered set. It determines the maximum possible difference along one of the coordinates. The maximum distance between two points on the cube is the diagonal of the cube; it is the sum of the basis vectors. Then the relative difference d between two vectors is defined as

$$d = \frac{\left|\vec{a} - \vec{b}\right|}{\left|\sum_{i=1}^{n} \vec{e}_i\right|}, \quad \text{where} \quad \begin{array}{l} \vec{e}_1 = (1;0;\ldots;0) \\ \vdots \\ \vec{e}_n = (0;0;\ldots;1) \end{array} \tag{4.17}$$

This difference value lies in the interval (0; 1) as can be clearly seen from the illustration for a three-deminational unit cube, given in Fig. 4.1. All vectors from gathered set (black points) are within the cube formed by the basis vectors. The diagonal of the cube is the maximum possible distance between any two vectors.

If for two vectors $a = (a_1,\ a_2, \ldots,\ a_{\mathrm{k}})$ and $b = (b_1,\ b_2, \ldots,\ b_{\mathrm{n}}) k < n$ then vector **a** can be transformed to n-dimensional space by adding complementary zero coordinates.

(4.17) is a more general version of the formula for separate numbers (4.14), which might be regarded as single-space vectors. Indeed, for real numbers a and b the relative difference according to (4.17) is

$$d = \frac{\sqrt{\left(\frac{a-b}{range}\right)^2}}{\sqrt{1}} = \frac{|a-b|}{range}$$

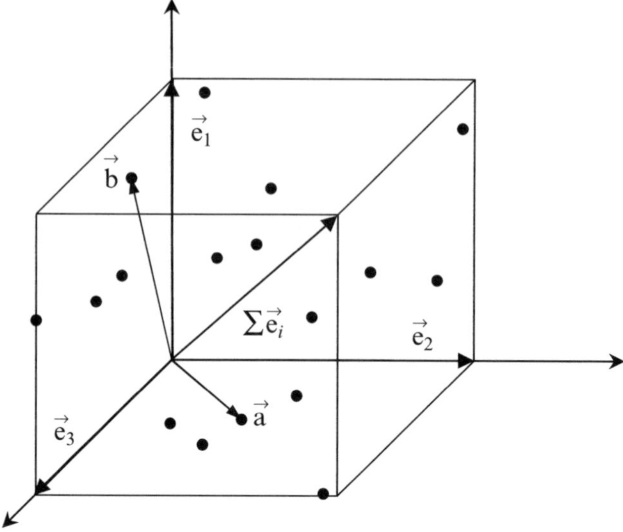

Fig. 4.1. Distances in a three-dimensional unit cube of all vectors of a gathered set

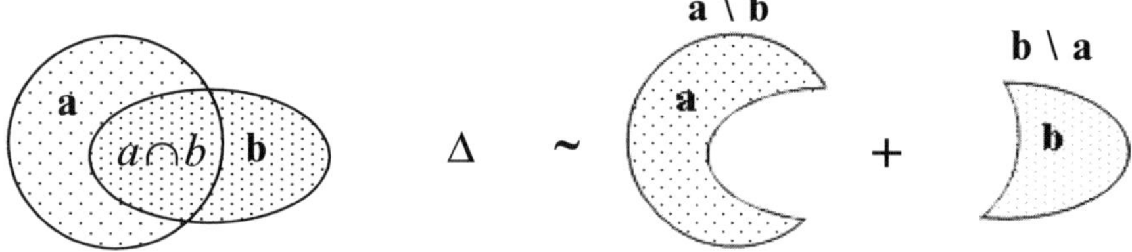

Fig. 4.2. Diagram of sets *a, b* and the determination of their difference

Sets

The values of corresponding features might be represented as sets. The difference value of such value-sets is determined by the number of elements in the sets which are not common, i.e. the difference for sets a and b equals the cardinal number of set $(a\backslash b) \cup (b\backslash a)$ – the sum of the difference of a from b and the difference of b from a (see Fig. 4.2).

Writing it in a more convenient way:

$$\Delta = |(a\backslash b) \cup (b\backslash a)| = |a \cup b| - |a \cap b| \tag{4.18}$$

The lower the number of common elements of the set the bigger the difference value. If the sets do not have common elements, then the difference is maximum, and it equals a sum of cardinalities of sets a and b, i.e. the cardinal number of the union of the sets.

Hence, the relative difference of two sets is determined as

$$d = \frac{|a \cup b| - |a \cap b|}{|a \cup b|} = 1 - \frac{|a \cap b|}{|a \cup b|} \tag{4.19}$$

Many types of data such as signs, symbols, and specific codes, cannot be compared quantitatively and require exact matching. The elements containing such data can also be regarded as sets including just one element. In the case of exact matching, the intersection coincides with the union of the sets and the relative difference equals 0; in other cases the intersection has no elements and the relative difference equals 1 [see (4.19)]. Summarizing for the elements that require exact matching the relative difference is determined as

$$\begin{matrix} a = \left\{ \overset{\frown}{a} \right\} \\ b = \left\{ \overset{\frown}{b} \right\} \end{matrix} \quad d = \begin{cases} 1, & \overset{\frown}{a} \neq \overset{\frown}{b} \\ 0, & \overset{\frown}{a} = \overset{\frown}{b} \end{cases} \tag{4.20}$$

Sequences

The values can be represented by ordered sets: sequences. When determining the difference value between two sequences it is important to take into account the position of an element in a sequence. The absolute difference of two sequences is calculated by counting the positions in which the elements of the sequences differ:

$$\Delta = \sum_{i=1}^{\max(|a|,|b|)} \mathit{diff}\,(a_i, b_i), \text{ where } \mathit{diff}\,(a_i, b_i) = \begin{cases} 0, & a_i = b_i \\ 1, & a_i \neq b_i \end{cases} \tag{4.21}$$

If the lengths of the sequences are not equal, then the *diff*-function gives 1 for non-existent elements of the smaller sequence. The maximum difference occurs when all elements in the sequences are different or the positions of identical elements are different. It equals to the number of elements in the biggest sequence. Hence, the relative difference is defined as follows:

$$d = \frac{\sum\limits_{i=1}^{\max(|a|,|b|)} \mathit{diff}\,(a_i, b_i)}{\max(|a|, |b|)} \tag{4.22}$$

The formula can be applied for sequences of symbols if the symbols cannot be aggregated in some codes or meaning words.

Graphs

If a value of feature a is described by both a set v and a relationship between the elements of set v, that means the subset e in the Cartesian product $v \times v$, then such a set should be considered together with their relationships and might be attributed to a class of graphs.

A graph is a pair of sets $a = (v, e)$, where v – not empty set, and e are a subset of all ordered or disordered pairs of the different elements of a. The difference of such structures relates to the notion of graph isomorphism.

Let $A_1 = [v_1, e_1], A_2 = [v_2, e_2]$ – two graphs. Assume there is a function $f : v_1 \rightarrow v_2$, that the following expressions are correct:

(1) if $x, y \in v_1,\ x \neq y$, then $f(x) \neq f(y)$;
(2) $\forall\ y \in v_2,\ \exists\ x \in v_1\ :\ f(x) = y$;
(3) if $(x, y) \in e_1$, then $(f(x), f(y)) \in e_2$;
(4) $\forall\ (p, q) \in e_2,\ \exists\ (x, y) \in e_1 :\ p = f(x),\ q = f(y)$.

Then f is an isomorphism of graphs $A_1,\ A_2$, and the graphs are isomorphous (Belskiy, 1979).

Before considering the determination of the difference value for isomorphous graphs and for non-isomorphous graphs one particular case will be described.

If the values of corresponding features in the entities are subgraphs of a common graph then the values can be considered just as pairs of sets.

Let a common graph be denoted as $\mathrm{C} = (\mathrm{U}, \mathrm{Y})$. For two values $a_1 = [v_1, e_1], a_2 = [v_2, e_2]$, where $v_1,\ v_2 \subseteq U;\ e_1,\ e_2 \subseteq Y$, the absolute difference is defined using (4.18) for sets:

$$\Delta = \Delta_v + \Delta_e = |v_1 \cup v_2| - |v_1 \cap v_2| + |e_1 \cup e_2| - |e_1 \cap e_2| \tag{4.23}$$

and the relative difference is:

$$d = \frac{\Delta}{\Delta_{\max}} = 1 - \frac{|v_1 \cap v_2| + |e_1 \cap e_2|}{|v_1 \cup v_2| + |e_1 \cup e_2|} \tag{4.24}$$

If two graphs $a_1 = [v_1, e_1], a_2 = [v_2, e_2]$ are isomorphous, i.e. there is a function f are defined above, then there is such a subset v^* of set v_1 that the following statement is correct: $\forall\ x \in v^*,\ \exists\ y \in v_2\ :\ f(x) = y$; and there is such a subset e^* of set e_1 that the statement $\forall\ (x, y) \in e^*,\ \exists (p, q) \in e_2\ :\ p = f(x),\ q = f(y)$ is correct. The subgraph $a^* = [v^*, e^*]$ of graph a_1 is a mapping of graph a_2 into graph a_1; it has the same topology as graph a_2 and therefore they are topologically identical. The value of the difference between graph a^* and a_1 equals the difference between a_1 and a_2. Since the graph a^* is a subgraph of graph a_1 the absolute difference is determined by (4.24) and the relative difference is calculated by (4.25).

If two graphs $a_1 = [v_1, e_1], a_2 = [v_2, e_2]$ are not isomorphous, then the first step is to find the biggest subgraph in one of the graph that is isomorphous with another graph. This subgraph $a^i = (v^i, e^i)$ can be mapped to the topologically identical subgraph $a^* = (v^*, e^*)$ in an isomorphous graph. The absolute difference between a^i and a_1 as well as the absolute difference between a^* and a_2 are defined by (4.24). Then the distances are summed. The maximum

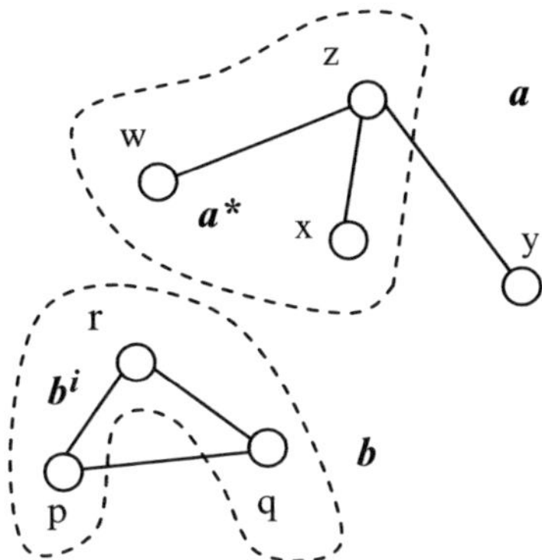

Fig. 4.3. Representation of graphs a and b

difference is when no subgraph which is isomorphous to one of the graphs is found; it equals to the sum of the cardinalities of the unions of the vertexes and edges of both graphs. The relative difference is then defined as follows:

$$d = \frac{\left|v_1 \cup v^i\right| + \left|e_1 \cup e^i\right| + \left|v_1 \cup v^*\right| + \left|e_1 \cup e^*\right| - \left|v_1 \cap v^i\right| - \left|e_1 \cap e^i\right| - \left|v_1 \cap v^*\right| - \left|e_1 \cap e^*\right|}{\left|v_1 \cup v_2\right| + \left|e_1 \cup e_2\right|} \tag{4.25}$$

Explanations. There are two graphs a and b, which are geometrically represented in Fig. 4.3. The biggest subgraph in b which is isomorphous with a is circled by dashed lines and denoted as b^i. Its mapping in graph a is graph a^*.
The absolute difference values are:

$$\Delta(b, b^i) = 3 - 3 + 3 - 2 = 1$$
$$\Delta(a, a^*) = 4 - 3 + 3 - 2 = 2$$

The relative difference is determined as:

$$d(a, b) = \frac{1 + 2}{6 + 7} = \frac{3}{13}$$

The most difficult problem in the described procedure is to identify the proper subgraph in one graph that is isomorphous with another graph. This problem requires utilization of suitable algorithms, which are not considered in this work.

Another particular case of graphs is when the values of corresponding features are dangling vertexes of a common tree.

Explanations. A tree is a bonded graph without circuits. A graph is bonded if any two of its vertexes are bonded, i.e. connected by a path. A path is a record of the vertexes or branches of graph that form a way from one vertex (the beginning of the path) to another vertex (the

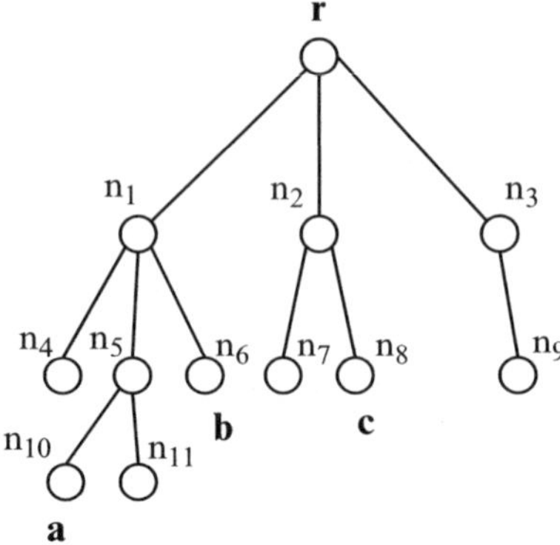

Fig. 4.4. An example of hierarchy

end of the path) on the graphical interpretation of the graph. A path without repeated branches is a chain. A chain, where the beginning and the end of a path is the same vertex, is a circuit. A dangling vertex has a degree equalling 1.

Often the tree is a certain hierarchy of elements, and the elements might be located in different branches and at different depths. An element can be described by the path from the root of the tree to the dangling vertex that corresponds to this element. For the tree of Fig. 4.4, the elements a, b and c are described as

$$a = \{r, n_1, n_5, n_{10}\};\ \ b = \{r, n_1, n_6\};\ \ c = \{r, n_2, n_8\}$$

The magnitude of difference for such elements is defined by the maximum difference of their paths: the maximum length of path from the first common node where their paths met, i.e.

$$\Delta = \max(|a \backslash b|\, , |b \backslash a|) \tag{4.26}$$

Or in a more convenient way

$$|a \backslash b| = |a| - |a \cap b|\, ; \quad |b \backslash a| = |b| - |a \cap b|$$
$$\Delta = \max(|a \backslash b|\, , |b \backslash a|) = \max(|a| - |a \cap b|\, , |b| - |a \cap b|) = \max(|a|\, , |b|) - |a \cap b| \tag{4.27}$$

The maximum difference is for elements that are located in different tree branches connecting only at the root. If the root of the tree is not included in the path description, then the maximum difference equals to the maximum length of one of the elements. Thus the relative difference is determined as

$$d = \frac{\max(|a|\, , |b|) - |a \cap b|}{\max(|a|\, , |b|)} = 1 - \frac{|a \cap b|}{\max(|a|\, , |b|)} \tag{4.28}$$

Explanations. For the example shown in the difference value for elements a and b is (the root of the tree is neglected):

$$d(a,b) = 1 - \frac{1}{\max(3,2)} = \frac{2}{3}$$

For the elements a and c the difference is 1, as their paths meet only in the root:

$$d(a,c) = 1 - \frac{0}{\max(3,2)} = 1$$

Qualitative Values

The qualitative values can be grouped to some categories. The difference between the values belonging to the same qualitative category is considered equal to 0. Otherwise, the difference between two qualitative values is determined by the number of other categories separating the values from each other on the qualitative scale. An integer value can be assigned to each qualitative category and the value of difference between two qualitative variables can be calculated by determination of the difference between these integer numbers according to (4.14). The maximum distance of 1 is between the highest and the lowest categories. The *range* coefficient corresponds to the highest integer number that is assigned to a qualitative category in the considered qualitative scale (if the categories are encoded starting from number 1).

4.2.4 Determination of Difference for Composite Values

In some cases the values of the features of an entity cannot be represented by basic data formats. They also cannot be represented as a new entity because they correspond to one property of an artefact. Such values require a special approach to determine the difference between them. A few types of composite values will now be considered.

Sets of Structured Elements

Often, the set in the description of an entity contains elements which are represented by a composite data structure. One of examples is a set, each element of which is a real number. When determining of the difference value between such composite sets the difference between the elements of sets must be counted. The question is how to find the corresponding elements in both sets. It can be done by comparing the difference values between different elements. The smaller the difference of two elements of the sets, the more similar they are, and they can be regarded as corresponding to each other. These elements build a closest pair.

Let us consider two composite sets a and b, for which a difference value is determined as a bipartite graph G. The elements of both sets are vertices

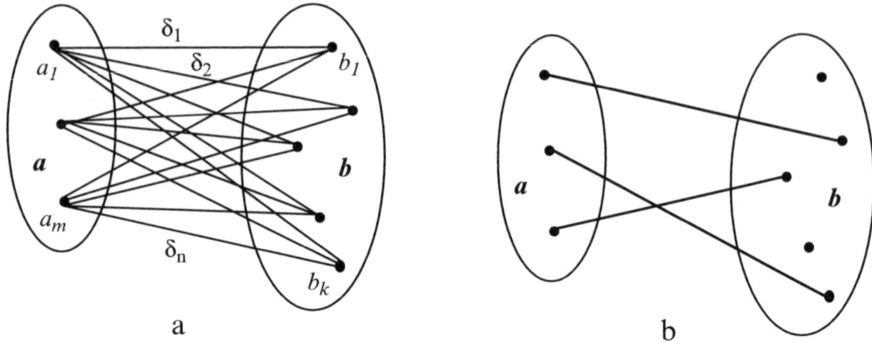

Fig. 4.5. Bipartite graph representation of two sets (**a**) complete covering (**b**) maximum matching

of the graph G, the edges of the graph are all possible pairs of elements of sets, where one part of a pair belongs to set a, and another part belongs to set b (Fig. 4.5a). A numeric value corresponds to each edge in the graph. This value $\boldsymbol{\delta}$ is the difference between two numeric elements to which the edge related. It is necessary to find a set of closest pairs that has minimum sum of differences of elements. In other words, the maximum matching of graph G with the minimum sum of edges' values $\boldsymbol{\delta}_i$ should be identified. Matching is a subgraph of graph G, any edge of which has no common vertex with other edges (e.g. Fig. 4.5b).

The absolute difference is then determined as follows (M is found matching):

$$\Delta = \sum_{i=i}^{\max(|a|,|b|)} \delta_i, \text{ where } \delta_\mathrm{i} = \begin{cases} d(a_i, b_j), & (a_i, b_j) \in M \\ 1, & b_i \notin M \text{ } or \text{ } a_i \notin M \end{cases} \tag{4.29}$$

The absolute difference value contains the differences between elements that belong to the maximum matching, if an element of one set has no match in another set (it happens when the sets are of different cardinalities) then the difference for such an element is assigned to 1.

The difference is of maximum value when the distances between all elements from matching equal 1. Then the relative difference for composite sets is defined as:

$$d(a,b) = \frac{\sum_{i=i}^{\max(|a|,|b|)} \delta_i}{\max(|a|,|b|)}, \text{ where } \delta_\mathrm{i} = \begin{cases} d(a_i, b_j), & (a_i, b_j) \in M \\ 1, & b_i \notin M \text{ } or \text{ } a_i \notin M \end{cases} \tag{4.30}$$

The determination of the difference value between elements of the sets is independent of the calculation of the difference between the composite sets; therefore the (4.30) and (4.31) are valid even if the elements of the sets are not numbers but vectors or other data types.

Hierarchies with Assigned Assessments

The description of a simple hierarchy and the determination of difference values for its elements has been described previously. However, in some situations the paths in the hierarchy cannot be regarded as equal even if they have the same depths. In such hierarchies the edges of the tree have numeric values assigned to show the disparity.

Let the values assigned to each edge in the hierarchy be denoted as v_i, where the index corresponds to an edge. An element of the hierarchy is a path – a subset of the edges of the hierarchy.

The difference values for such a hierarchy are determined in a way similar to those for the simple hierarchy described above, except one circumstance: when counting the difference between the paths each element of the paths is considered as of length v_i. Therefore, the absolute and relative differences for two paths a and b are defined as:

$$\Delta = \max(\sum_a v_i, \sum_b v_j) - \sum_{a \cap b} v_k, \quad d = 1 - \frac{\sum_{a \cap b} v_k}{\max(\sum_a v_i, \sum_b v_j)} \tag{4.31}$$

Explanations. A hierarchy is represented as a tree, given in Fig. 4.6. The values assigned to the edges are selected in such way that a path from the root to a dangling vertex always has length 1. The difference value for elements a and b is:

$$d(a, b) = 1 - \frac{0.3}{1} = 0.7$$

For the elements a and c the difference is 1, as their paths meet only in the root:

$$d(a, c) = 1 - \frac{0}{1} = 1$$

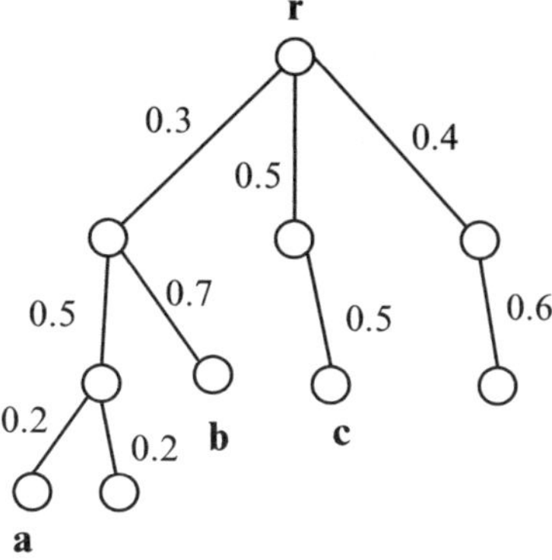

Fig. 4.6. A hierarchy with assigned values

It can be seen that for the hierarchy the length of union of the paths of two elements shows their degree of similarity. Such a hierarchy is called a similarity tree.

4.3 Concept of Adaptation

The adaptation procedure is based on the assumption that not only the most similar case can be used but also a set of cases located nearby the current problem in the problem space. The key assumption that a similar problem has a similar solution means that solutions of similar problems are located nearby each other. The distances between a new solution and solutions of the most similar problems must correspond to the distances between the current problem and the similar problems under consideration.

Since the solutions are also described by a set of information entities, the computation of distance is also based on the general similarity concept.

4.3.1 Foundations of Adaptation Method

Let us introduce the necessary symbols. The current problem to be solved is denoted as entity N, j-past problem and its solution has symbols P_j and S_j correspondingly (entities also), and a new solution being generated is denoted as entity C.

According to (4.12) the degree of similarity and degree of difference are supplementary notions. A small difference corresponds to a great similarity and vice versa. The distances between problems in the problem space and between solutions in the solution space can be characterized by degrees of similarity.

A new solution for the current problem is created based on a group of solutions of the most similar problems. A set of cases, which represents the minimum neighbourhood of the current problem, is denoted as L. The relative distances between the current problem and the problems from neighbourhood L are transferred to the solution space. The intersection of distance segments starting from solutions from neighbourhood L gives the point of a new solution (see Fig. 4.7).

Thus, the degree of similarity between a problem from the neighbourhood L and the current problem N must be equal to the value of the degree of similarity between a solution of that problem and the created solution. In the framework of the adaptation concepts this condition is called the condition of conservation of similarity and defined as:

$$f_j(C) = sim(S_j, C) - sim(P_j, N) = 0; \;\; (P_j, S_j) \in L \tag{4.32}$$

This is a function of the created solution C, since the problems of the neighbourhood L and their solutions are fixed and selected during retrieval.

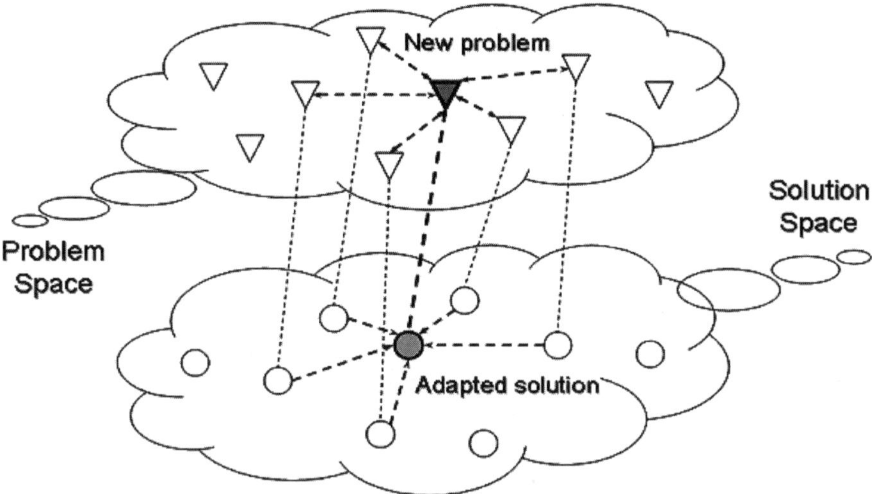

Fig. 4.7. Finding an adapted solution based on distances in the problem space

The goal is to find a solution C, which has the condition of conservation of similarity satisfied for all cases from the neighbourhood L.

The adaptation task is introduced as the minimization of the function:

$$F(C) = \sum_{j \in L} |f_j(C)| = \sum_{j \in L} |sim(S_j, C) - sim(P_j, N)| \tag{4.33}$$

where index j represents j-case.

Ideally, this adaptation function has to equal 0, where the conditions of conservation of similarity are exactly satisfied. But practically, it is a complicated task to find an ideal solution when the solution is represented by a large variety of design variables. A satisfactory tolerance is acceptable in most situations.

4.3.2 Scaling of Solution Space

The statement of adaptation described above implies equality of distance measures in the problem and solution spaces. It means that an elementary unit vector in the solution space is equivalent to such a vector in the problem space. However, this is rarely true. Coefficients should be applied that correct the differences in scales of both the spaces. These coefficients must be addressed to the solutions as the problems are heavily used in retrieval and the scale of the problem space better remains the same.

In order to take into account differences in the scales of the solution and problem spaces, scaling factors as weights for similarity in solution spaces are introduced.

It can be stated that the degree of similarity between two different problems must equal the degree of similarity between their solutions. In the framework of the adaptation concept it is called a similarity equivalence condition.

For two arbitrary solutions $S_i = (s_1^i, s_2^i, \ldots, s_k^i)$ and $S_j = (s_1^j, s_2^j, \ldots, s_k^j)$ the similarity equivalence condition is defined as:

$$\varphi_{i,j}(\vec{w}) = \frac{\sum_k^{\dim \vec{w}} w_k \cdot sim(s_k^i, s_k^j)}{\sum_k^{\dim \vec{w}} w_k} - sim(P_i, P_j) = 0 \tag{4.34}$$

where $\dim \vec{w}$ is the dimension of vector of scaling factors w_i.

For the close neighbourhood of a solution S_i with cardinality N_i the following statement should be correct:

$$\Phi_i(\vec{w}) = \sum_j^{N_i} \varphi_{i,j}(\vec{w}); \qquad \Phi_i(\vec{w}) = 0; \tag{4.35}$$

The use of a close neighbourhood can be explained by the fact that the ratio of scales between the solution and problem spaces might not be the same in different distant segments of these spaces. In contrast, the small neighbourhood shows a stable ratio between scales.

The number of neighbourhoods is selected to be equal to a dimension of vector $\vec{w}$, and one solution can be a member of different neighbourhoods.

The task of scaling the solution space is stated as follows:

$$\begin{aligned} &\Re : \{\Phi_i(\vec{w}) = 0; \quad i = 1..\dim \vec{w} \\ &\vec{w} = \arg \Re \end{aligned} \tag{4.36}$$

The vector of scaling factors can be found by solving the given system of algebraic equations.

4.3.3 Solution of a Adaptation Task

Usually, the three most similar cases are used. That is, the cardinality of neighbourhood for the adaptation task is 3. It may vary from task to task. The initial values of the features of new solution are copied from the most similar case data. The design parameters of a new solution are changed to reach the minimum of function F.

Because the design parameter may be of various type of data representation (combination of sets, graphs, vectors etc.), standard optimisation methods are not suitable. There is a need for a method, search strategy which is not dependent on variables. The genetic algorithm is perfectly suited to dealing with heterogeneous variables representation since it transforms all variables into an internal representation – a genome. Another advantage of this technique by its random nature in the search process; hence it allows a novel candidate to be obtained that could be similar to a retrieved solution but does not necessarily conform to it. For these reasons, the genetic algorithm has been selected as the global optimisation method. The objective function for the algorithm is the adaptation function $F(C)$, according to (4.34).

4.3.4 Description of Genetic Algorithm

The genetic algorithm is at the core of evolutionary methods. Evolutionary algorithms are stochastic search methods that mimic the metaphor of natural biological evolution. Evolutionary algorithms operate on a population of potential solutions applying the principle of survival of the fittest to produce better and better approximations to a solution. At each generation, a new set of approximations is created by the process of selecting individuals according to their level of fitness in the problem domain and breeding them together using operators borrowed from natural genetics. This process leads to the evolution of populations of individuals that are better suited to their environment than the individuals that they were created from, just as in natural adaptation.

Evolutionary algorithms model natural processes, such as selection, recombination, mutation, migration, locality and neighbourhood. Figure 4.8 shows the structure of a simple genetic algorithm. Evolutionary algorithms work on populations of individuals instead of single solutions. In this way the search is performed in a parallel manner.

At the beginning of the computation a number of individuals (the population) are randomly initialized. The objective function is then evaluated for these individuals. The evaluation function is used to measure the genome

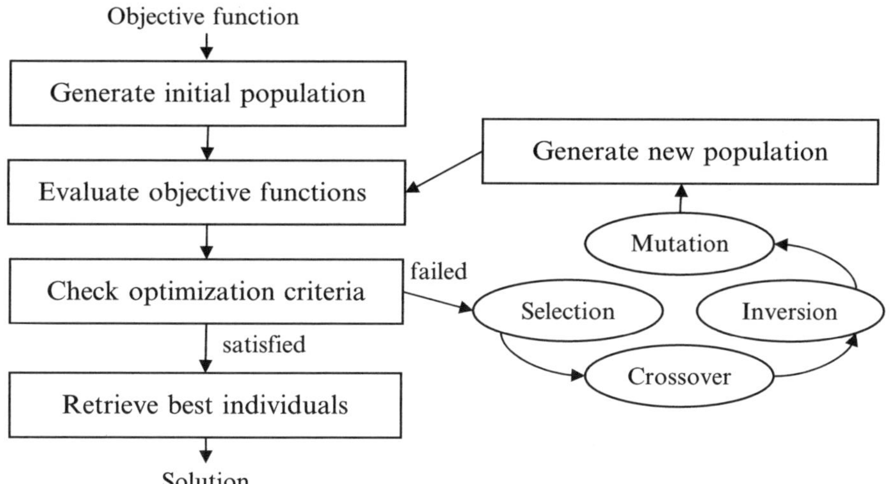

Fig. 4.8. Structure of a genetic algorithm

performance, or fitness, for the problem to be solved. If the optimization criteria are not met, the creation of a new generation starts. Individuals are selected according to their fitness for the production of offspring. Parents are recombined to produce offspring. All offspring will be mutated with a certain probability. The fitness of the offspring is then computed. The offspring are inserted into the population replacing the parents, producing a new generation.

As a result, after a number of successive reproductions, the less fit genomes become extinct, while those best able to survive gradually come to dominate the population. This process is performed until the optimization criteria are reached.

The genetic algorithm is an iterative process. Each iteration is called a generation.

The physical variables from the domain are represented in evolutionary methods in the form of chromosomes. Each variable is encoded by a fragment of a chromosome – a gene. Each letter in a chromosome is usually of binary nature – it takes the value 0 or 1. Any gene of a chromosome that encodes a variable has a constant number of letters (see Fig. 4.9). In principle, the chromosome can be encoded/decoded using conventional binary code, but usually the Gray code is applied for coding. Table 4.1 shows the difference between Gray and binary codes, and contains a conversion expression to get a real value for each value of the code (the range of real variables is represented by an interval $(a_i,\ b_i)$). The Gray code has advantages over binary code because getting the following value of the code requires only one operation of inversion of a letter in the preceding code value (see Table 4.2). In order to get a value with a decimal shift 5 in the Gray code, only the last letter of the value with shift 4 is inverted, whereas in binary code there is a need for two operations of inversion. This characteristic of the Gray code provides finer tuning during optimization.

The operations performed on chromosomes are: mutation, inversion and crossover. The essence of the operations can clearly be seen from Fig. 4.10. The position of application of an operation is selected randomly.

The crossover operator exchanges the chromosome parts. As a result, two new offspring are created. If a pair of chromosomes does not cross over, then chromosome cloning takes place, and the offspring are created as exact copies of each parent.

Mutation, which is rare in nature, represents a change in the gene. It may lead to a significant improvement in fitness, but more often has rather harmful

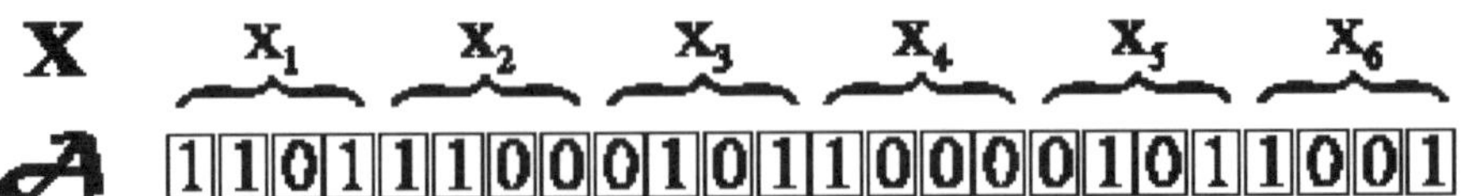

Fig. 4.9. Chromosome representation of variables

Table 4.2. Gray code representation and conversion to real value

Gray code	Binary code	Decimal shift	Conversion to real value
0000	0000	0	a_i
0001	0001	1	$a_i + 1(b_i - a_i)/15$
0011	0010	2	$a_i + 2(b_i - a_i)/15$
0010	0011	3	$a_i + 3(b_i - a_i)/15$
0110	0100	4	$a_i + 4(b_i - a_i)/15$
0111	0101	5	$a_i + 5(b_i - a_i)/15$
0101	0110	6	$a_i + 6(b_i - a_i)/15$
0100	0111	7	$a_i + 7(b_i - a_i)/15$
1100	1000	8	$a_i + 8(b_i - a_i)/15$
1101	1001	9	$a_i + 9(b_i - a_i)/15$
1111	1010	10	$a_i + 10(b_i - a_i)/15$
1110	1011	11	$a_i + 11(b_i - a_i)/15$
1010	1100	12	$a_i + 12(b_i - a_i)/15$
1011	1101	13	$a_i + 13(b_i - a_i)/15$
1001	1110	14	$a_i + 14(b_i - a_i)/15$
1000	1111	15	b_i

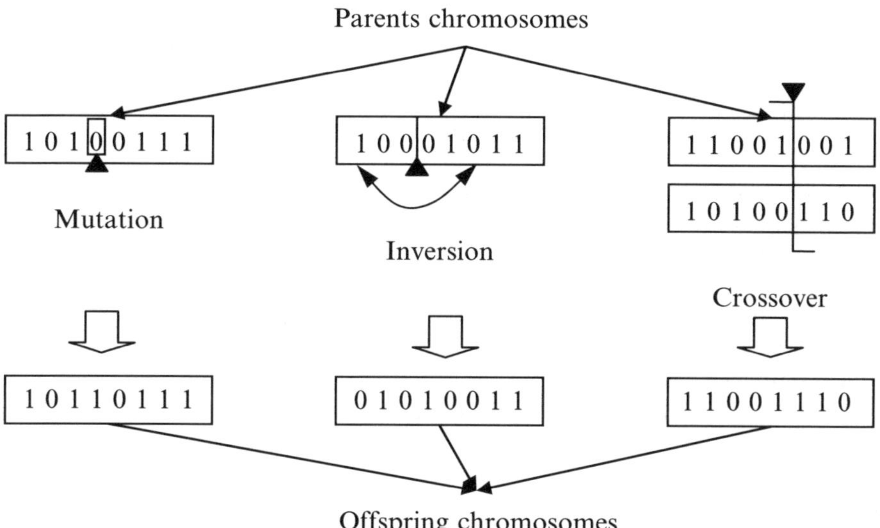

Fig. 4.10. Operations of a genetic algorithm

results. The role of mutation is to provide a guarantee that the search is not trapped in the local optimum.

Another important operation in a genetic algorithm, as can be seen in Fig. 4.8, is selection. During selection only the best individuals are chosen to be the parents for a new population. The first step is fitness assignment. Each individual in the selection pool receives a reproduction probability depending on the own objective value and the objective value of all other individuals in the selection pool. This fitness is afterwards used for the actual selection step.

Summarizing, the outline of the process of the genetic algorithm is as follows:

1. Generate random population of n chromosomes (suitable solution for the problems).
2. Evaluate the fitness F(x) of each chromosome x in the population.
3. Create a new population by repeating the following steps until the new population is complete:
   a. Select two parent chromosomes from a population according to their fitness (the better fitness, the greater chance of being selected).
   b. With a crossover probability cross over the parents to form new offspring. If no crossover is performed, the offspring is the exact copy of one of the parents.
   c. With an inversion probability the inversion operation for offspring at random position is performed.
   d. With a mutation probability mutate new offspring at random position.
   e. Place the new offspring in the new population.
4. Use the new generated population for a further run of the algorithm.
5. If the end condition is satisfied, stop, and return the best solution in the current population.
6. Return to step 2.

As can be seen from the outline presented above, the main parameters of the genetic algorithm are: crossover probability (should be about 80–95%), inversion probability (around 0.1%), mutation probability (best rates are about 0.5–1%), and population size (sizes 30–50 are reported as the best).

The search space of design cases is too large that it could be expected to exhaustively consider all possible potential candidate solutions. It is common to use a criterion based on the designer's intention such as the number of adaptation cycles executed. Another criterion is that the objective function for an obtained population in the genetic algorithm is less than a predetermined constant.

4.4 Summary

The design problems in chemical engineering are often quite difficult to represent as a well-structured list of features of one or two data types. The representation of design cases requires various models because design content involves topological, geometric, and physical properties and relations between them. Many problems in chemical engineering are very large and complex, the problem description is often incomplete and uncertain. The proposed general similarity concept is able to cope with cases that have different structure representation in the case base and contain the features expressed in different formats. The developed concept of adaptation is task independent and relies on the general similarity concept. The adaptation method requires an optimization procedure. The genetic algorithm is perfectly suited to dealing with heterogeneous variables representation since it transforms all variables into genome code.

5

Case-Based Reasoning Environment – Cabareen

5.1 Introduction

The process of the development of assistant tools for specific design problems evolves to the creation of general computer system – a case-based reasoning environment, CaBaReEn. The environment implements almost all the steps of the case-based design support methodology. It contains a set of functional elements which provides the basis for any application to support the design problems. All problems described further can be supported with the CaBaReEn system.

CABAREEN – is a software product designed to be used as a supportive and implementing tool for an application of case-based design supporting method described in this work. It allows the user to create a case-based reasoning tool for specific application. The user can make its application using many features that implement most activities of design supporting method. Among other features there are abilities to link specific interface for data management for concrete application, running script of commands, supporting several data formats, registration of new utilities to facilitate the process.

This tool is not a ready case-based reasoning system but an environment where a specific application of case-base design supporting paradigm can be developed. It provides a set of functions facilitating the development process. Such procedures as compilation of case base from different sources, building case structure from a file of description of properties, comparing the cases based on set of embedded and external similarity functions and many others are implemented in the environment.

The case-based reasoning environment is purposed to facilitate the process of development of a decision supporting system utilizing case-based reasoning technique and to reduce the required development time.

Y. Avramenko and A. Kraslawski: *Case-Based Design*, Studies in Computational Intelligence (SCI) **87**, 99–105 (2008)
www.springerlink.com

5.2 The Core of the Environment

The functions of the environment are realized in commands. The environment is represented as a standard application for operation system which is able to execute the commands implementing certain functions of the environment. A command for the environment is introduced via command line located on the bottom of main window. The results of execution of a command appear in black screen above the command line (see Fig. 5.1). This area serves to display the messages of the environment such as error report or status report. The command line and message screen are formed the main window of the environment but it can be shaded (sent to back) by another windows of the environment if the user expands them to maximum size. Pressing Ctrl+Tab combination of keys switches between windows of the environment.

The environment can be used as a simple text editor where the tool bar on the top of the window contains buttons for operations with files (new, open and save) and clipboard (cut, copy and paste). The text files are highly used during working in the environment because scripts of commands, data source descriptions and adjusting parameters are stored in text files.

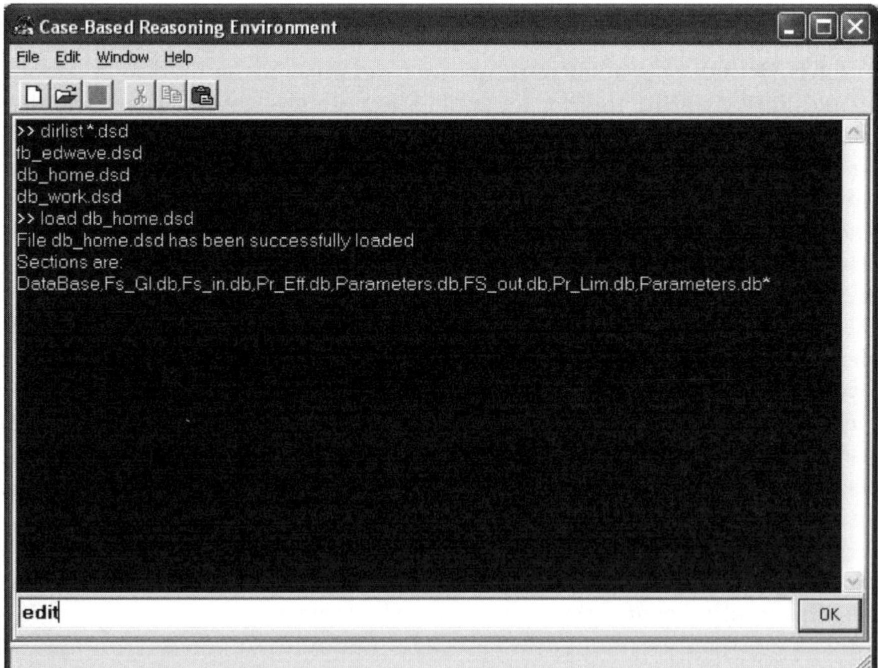

Fig. 5.1. Main screen of case-based reasoning environment with command line

The new text file can be created by pressing the first button on the tool bar with blank sheet image. Next button opens existing file from a given location (the file open dialogue appears to help finding a location of a file). Last button of the triplet is able to save a file under editing. The same operations can be done using corresponding actions in menu File. The actions from menu Edit are able to cut, copy or paste a piece of text from/to clipboard. These actions can be executed by means of corresponding buttons on the tool bar. Window actions rearrange opened windows of the environment; they are accessible from menu Window.

A command to be executed is typed in the command line with supplementing parameters and then OK button on the right or key Enter on the keyboard are pressed to execute the command. The list of available commands with description of its function is given in the Appendix I. Many commands require additional key and parameters; often, parameters can be omitted, and keys are not necessary part of a command. A key is separated from a command by space and slash ('/'). Parameters are always placed after all keys and separated from them by at least one space.

The full list of commands of the environment is given in the Appendix.

Many commands get the necessary data from files of specific format. Case base is complied in one or several files of format CML – case mark-up language, which is similar to XML format.

5.3 Links to the Environment

The commands implement most of functions required to prepare specific case-based reasoning application. However, the use of commands and specific files is not always convenient and requires some efforts. The use of different extensions, which are linked to the core of the environment, makes the environment a powerful tool.

Five types of facilities can be links to CABAREEN (Fig. 5.2).

Drivers provide the ability to work and understand various formats of data which serve as data source in creating case base. They are linked as dynamic link libraries. The environment includes Borland Database Engine (BDE) that allows it to work with most of popular database formats. Also, it contains drivers of internal format for various files of specifications, and case mark-up language (CML) format.

Utilities facilitate the work with data and commands of the environment. They perform various functions: navigation on data files and supporting files of the environments (e.g. utility 'sked'), navigation on fields of database (utility 'tabed'), adjusting parameters of embedded algorithm (e.g. utility 'ga') and so on. In the current version of the environment the utilities are registered in the system during compilation, therefore new utilities cannot be added dynamically. In this case utilities are windows-based extensions of commands. Utilities might also have parameters.

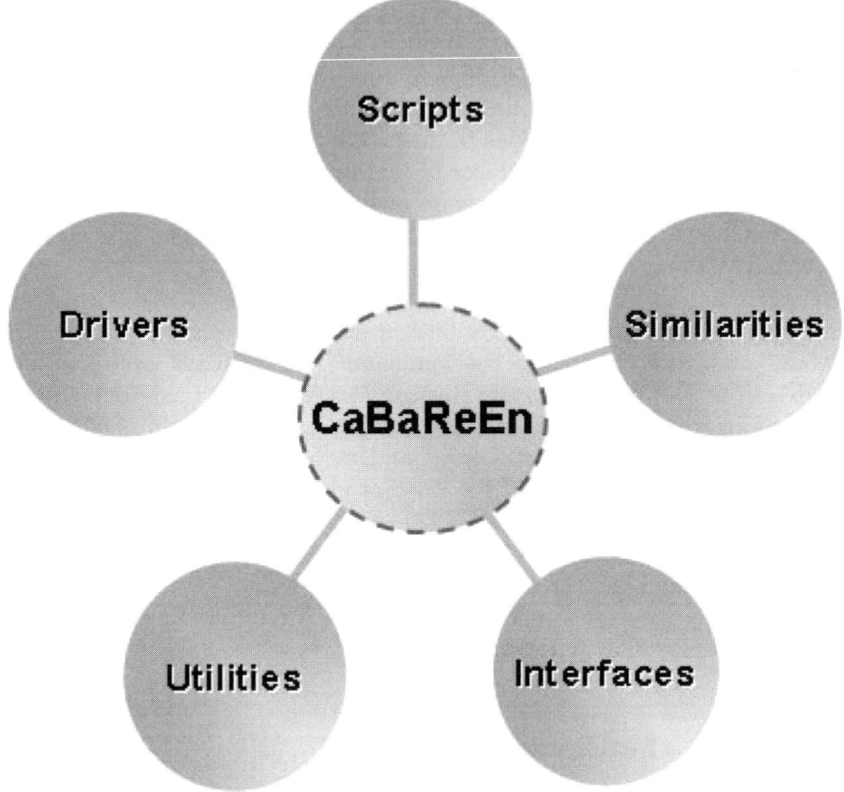

Fig. 5.2. Variants of links to CABAREEN

To compensate this limitation, interfaces can be linked to the environment. Interfaces are stand-alone application of operation system (like CABAREEN) but they can interact with core of the environment by means of messages and specific files.

The environment contains the similarity measurements (difference measurements) for basic types of data formats. However, real design applications very often require the specific similarity measurements for composite values. Additional similarity measurement can be linked to the system as applet in Gentee language (Gentee Inc.). More details about Gentee language can be found in http://www.gentee.com.

Scripts are lists of commands to be executed which are stored in text files. They are used to avoid constant typing repeated commands. Actually, the concrete application of the case-based design method is implemented via scripts in the environment. A script can manage with data stream and open necessary interface windows. An example of a script is given below.

```
-------------------------------------------
‘ Testing tables creation and navigate on
‘ data in database at work computer
load db_work.dsd
% sked
pause
dbinit
crtables/d short
setrelt
datacnt
% tabed
pause
% tabed c
% tabed 2
pause
run DBNavigation.exe
‘ Testing creation of the case base
load cb_str.csd
cbinit
crcb/db
run CBView.exe
dbclose
quit
-------------------------------------------
```

Thus, drivers are dynamic link libraries (DLL), utilities can be linked as Active X components during compilation, interfaces are separate executable files, similarities are linked as Gentee applets, and scripts are sequences of commands of the case-based reasoning environment.

5.4 Work with Databases

The work with relational databases is provided with commands `dbinit, crtables, setrelt, datacnt`, and `dbclose`.

Initialization of database, creation of objects for tables and setting of relation between tables require file with specifications of database – data source description file (DSD-file). The file includes location of database and tables descriptions.

The content of the file is divided on sections. A section begins with their name placed in brackets. One section describes entire database, other sections represents links and specifications of tables.

Section `DataBase` involves keys `Path`, representing full path to the folder where the database is located, and `Tables`, presenting list of tables, framed by quotation marks and separated by commas. If a table have links with more than one other tables from the database then it placed in additional key – `DuplicatedTables` – with specific symbol in the end of its name (symbol is '*').

Next sections must have titles correspond to tables' names from list presented in the key `Tables` of the section `DataBase`. A section of a table may have keys `MasterTable`, `MasterField` and `IndexField`, which contains name of linked tables, linked field and its own field, which serves as index for a link, correspondingly.

A part of DSD file, for example, is presented below.

```
(DataBase)
Path=D:\Inprise\DataBase storage\WaamDB_New\
Tables="Fs_GI.db","Fs_in.db","Pr_Eff.db","Parameters.db",
"FS_out.db","Pr_Lim.db"
DuplicatedTables="Parameters.db*"

(Fs_in.db)
MasterTable=Fs_GI.db
MasterField=ID_Key
IndexField=ID

(Pr_Eff.db)
MasterTable=Fs_GI.db
MasterField=ID_Key
IndexField=ID_Device

(Parameters.db)
MasterTable=Pr_Eff.db
MasterField=ID_Parametr
IndexField=ID_Parametr
...
```

5.5 Interfaces

With help of interfaces it is possible to manage convenient data input or parameters corrections for specific case-based reasoning application. Usually, it is graphical input window which dialogues with the user (see Fig. 5.3). All aspects of specific design application is better to represent using interfaces. The use of interfaces is not necessary because all data can be introduced into the environment vie text files of certain formats. However, in order to

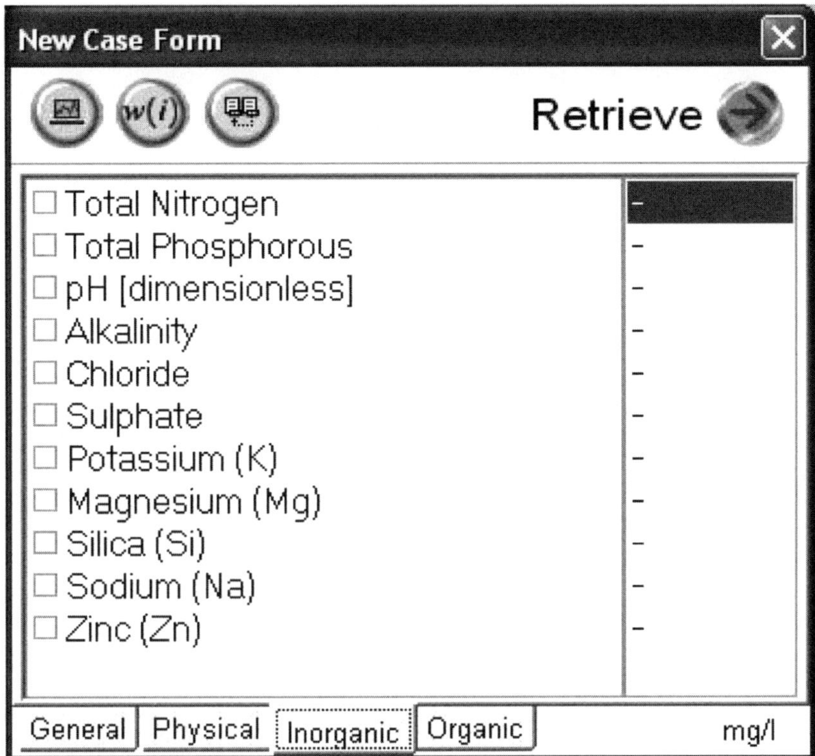

Fig. 5.3. Example of interface for introducing new problem

facilitate the input process or other functions of CBR process, to make specific application more attractive and convenient, interfaces are utilized.

An interface can be launched by using command `run` (e.g. `run intCS.exe`). Interface can store all related data into specific file, which then can be linked to the environment using command `load` (e.g. `load newcase.cs`).

The more examples of interfaces could be found in the following chapters where the specific applications to the design problems are considered.

Part III

Application to Support of Design Process

The examples of the application of the developed methodology do not represent the overall design process. Instead, only a crucial stage of each application is considered. The method described in the previous part, of course, can be applied to support the overall design process if the necessary design experiences are recorded, but in order to demonstrate the variety of applications the strategy of supporting one key stage was chosen.

Assistant tools implementing the case-based design methodology have been created to support the design of food ingredients (Product design stage), synthesis of a wastewater treatment sequence (Conceptual design), the construction of a model for distillation synthesis (Conceptual design phase), and the selection of column internals for reactive distillation (Equipment design). The tools serve to assist the design engineers rather than replace them. Engineer intervention is required to generate or evaluate a proper solution. Moreover, there may be a need for external validation of a proposed solution (using computational fluid dynamics, or a simulation of the model). The assistant tools are intended to reduce the development time and propose preliminary solutions to design problems. All problems described further can be supported with the CaBaReEn system.

6

Product Design: Food Product Formulation

The design of various food products focuses on identification of the structure and composition of food ingredients that have the desired characteristics. A thorough understanding of the functions and properties of the various ingredients is the basic key to formulating for the desired attributes. The revealing of structural properties of a food product is the main task of food formulation problem. Designing a product based on fats and oils blends is one of such tasks.

6.1 Introduction

Fats and oils are key functional ingredients in a large variety of prepared food. They are used in the development of shortening, margarine, and liquid oil product. Fats and oils have found utility because of their unique properties. These ingredients are used to add flavour, lubricity, texture and satiety to food. They are the highest energy source of the three basic nutrients (carbohydrates, proteins, and fats) and many contain fatty acids essential for health that are not produced by the human body.

Successful development of food products relies on effective use of the different functional properties of the available fats and oils and manipulation of the fat blend to satisfy the prepared food's requirements. The chemical composition defines the characteristics of the individual fat or oil, which in turn determines the suitability of this ingredient in various processes and application (O'Brien, 2004).

Application development of fat and oil products begins with identification of the key functional attributes that the final product is expected to provide. The important functional attributes which are considered for product development, are: lubricity, structure, clarity, consistency, plasticity, emulsification, creaming property, spreadability, aeration, hardness, freeze stability, flavour (odour, taste and mouth feel) and flavour stability. The product functionality

Y. Avramenko and A. Kraslawski: *Case-Based Design*, Studies in Computational Intelligence (SCI) **87**, 109–116 (2008)
www.springerlink.com

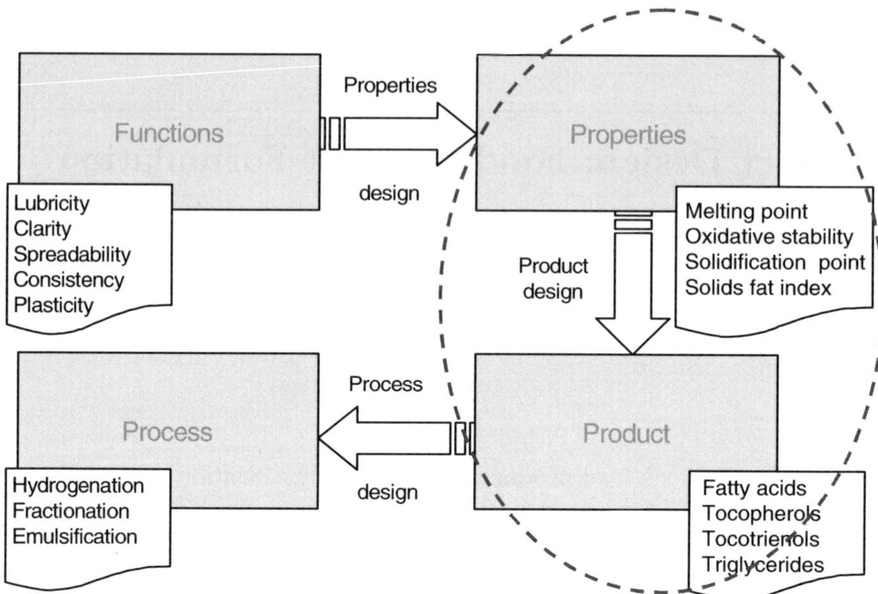

Fig. 6.1. Food product formulation stages

can usually be translated into analytical measurements and physical properties. For example, mouth feel and flavour release can be controlled by the melting properties and oxidative stability. The solids fat index (SFI) curve characterizes the consistency and spreadability of the product. This is the task of properties design of food ingredient development.

The objective of the product design is to identify the structural properties most likely to produce the intended functionality. The focus is on middle part of overall product development process. While the required physical and chemical properties of the product under development is determined in the properties design stage using as input intended functions, the structural properties (composition, solid–liquid distribution) are found in the considered product design stage (see Fig. 6.1) In order to achieve the objective, historical knowledge is used.

6.2 Database of Fats and Oils Properties

The database of physical properties and compositions of nature and genetically modified oils and fat blends has been created based on materials of book (O'Brien, 2004) and other sources. Materials have been gathered over the past 40 years from patents, trade journals, scientific journals, and reference books.

The list of fats and oils blends which are included to the database is presented in the Table 6.1. As stated before, understanding the functions and

Table 6.1. List of fats and oils collected in the database

Natural	Genetically modified	Natural	Genetically modified
Canola	Present	Olive	–
Coconut	–	Palm	Present
Corn	Present	Palm kernel	–
Cottonseed	Present	Peanut	Present
High-oleic safflower	–	Safflower	–
High-oleic sunflower	–	Soybean	Present
Lard	–	Sunflower	–
Milk fat	–	Tallow	–

properties of a shortening, oil, margarine or other fats-based product is a key element of proper usage and product formulation. Therefore, the database contains also characteristics of several groups of food products: baking shortening, frying shortening, household shortening, margarines and mayonnaise, dairy analog shortenings.

6.3 Case Representation of Fats and Oils Products

The chemical and physical properties of fats and oils are largely determined by fatty acids that they contain and their position within the triacylglycerol molecule. Chemically, all fats and oils are esters of glycerine and fatty acids; nevertheless, the physical properties of natural fats and oils vary widely because the proportions of fatty acids vary over wide ranges, and the triacylglycerol structures vary for each individual oil and fat.

A case description contains composition, physical and chemical properties as well analytical test results of fats and oils blends. The features of the fats and oils have been divided into four categories:

- Flavour, which is characterised by oxidative stability;
- Physical properties, such as melting point, refractive index, relative density, specific gravity etc.
- Textural properties, determined by crystal habit, solidification point, SFI etc.;
- Structural properties, which are defined by fatty acid composition, unsaponifiable number, tocotrienol content etc.

The oxidative stability is represented by separate entity because it is complex characteristic and some of its components might not be available. All physical properties are either simple number or vector types (like Refractive density, which is measured at certain temperature and the temperature of measurement is not always same for different oils). Textural features are of various types of representation. Crystal habit, for example, represents the type of crystal (β or β'), which is a logical matter. Solid Fat Index curve is represented by a set, each element of which is a vector in temperature-index space.

Most of the structural properties are vectors. Only the composition of fatty acids is organized as the set where an element of set indicates the concentration of an acid into the blend. The list of fatty acids is not the same for all oils. Some kinds of fatty acids could not be presented in a number of oils. Therefore, the set has variable number of element for each fat or oil record.

Flavour, physical and textural properties build the problem part. Its set of entities is given in Table 6.2. The solution part includes the four features: fatty acids, tocopherols, tocotrienols, and triglycerides (Table 6.3).

Table 6.2. List of entities in the problem description

Feature	Type
Cold test	Numeric
Cloud point	Number
Crystal habit	Logical
Melting point	Numeric
Oxidative stability	Entity
AOM test	Numeric
Inherent OS	Numeric
Iodine value	Numeric
Refractive index	Vector = (value, temperature)
Relative density	Vector = (value, temperature)
Saponification	Vector = (saponification, unsaponifable number)
Solidification point	Number
Solid Fat Index	Entity = set of vectors Each vector = (temperature, %)
Specific gravity	Number
Titer	Number
Wax	Number

Table 6.3. List of entities in the solution description

Feature	Type
Triglycerides	Vector = (trisaturated, disaturated, monosaturated, triunsaturated), %
Tocotrienols	Vector (α-, β-, γ-, δ-), ppm
Fatty acids	Entity = set of numbers (%)
Tocopherols	Vector (α-, β-, γ-, δ-), ppm

6.4 Similarity Determination

The functionality of the product in development translated to physical properties and test results is used to identify the oils with most similar properties. The similarity is calculated according to General similarity concept.

The measurement of similarity is depended of data type of a feature. Most of the features are of numeric type (single number or vector). There are a few features of composite value (Solid fat index, Fatty acids content). These are examples of a set of composite elements as it was described in Chap. 4.

The oxidative stability and the Solid Fat Index are represented as separate entities. One or two features of the oxidative stability might not be available but that fact is taken into account during similarity determination.

The Solid Fat Index curve is represented as a set, each element of which is a vector in temperature-index space. The similarity between to SFI values is determined as follows. First, the closest points of indexes are identified. It means the set of best matched points (temperature and index) according to the paired similarity is created. For example, if in one case the SFI curve starts from 21°C, next point is 35°C, but in another case the curve starts only from 41°C, the best matched points after calculation of vectors similarity will be 35–41°C, but the point 21°C of first case will not have match in second case. Next, the overall similarity of this entity is calculated based on best matching found.

The similarity value between new problem and past cases are determined with taking into account weights assigned for every feature. The weights were determined by learning of algorithm on test set of existing oils.

6.5 Computer Assistant for Support of Food Product Formulation

The computer assistant has been created based on the case-based reasoning environment, CaBaReEn. The overall process of supporting food product formulation task is realized as a script of CaBaReEn. Modules and interfaces have been developed to facilitate the new problem introduction, the management of the adaptation procedure, and control and validation of the created solution. The example of an interface developed for the introduction of a new problem is shown in Fig. 6.2.

The information describing the design cases is stored in a database in which a record is represented as an XML-text. A separate tool manages the database of fat and oil properties. The tool works with internal system data format based on XML. There is an ability to add data about new oils or new final food products. The structure of the assistant is shown in Fig. 6.3.

The case base is created from information sources during running of the system based on the given case representation that is produced depending on the task goal – I. A case is constructed from information entities found in

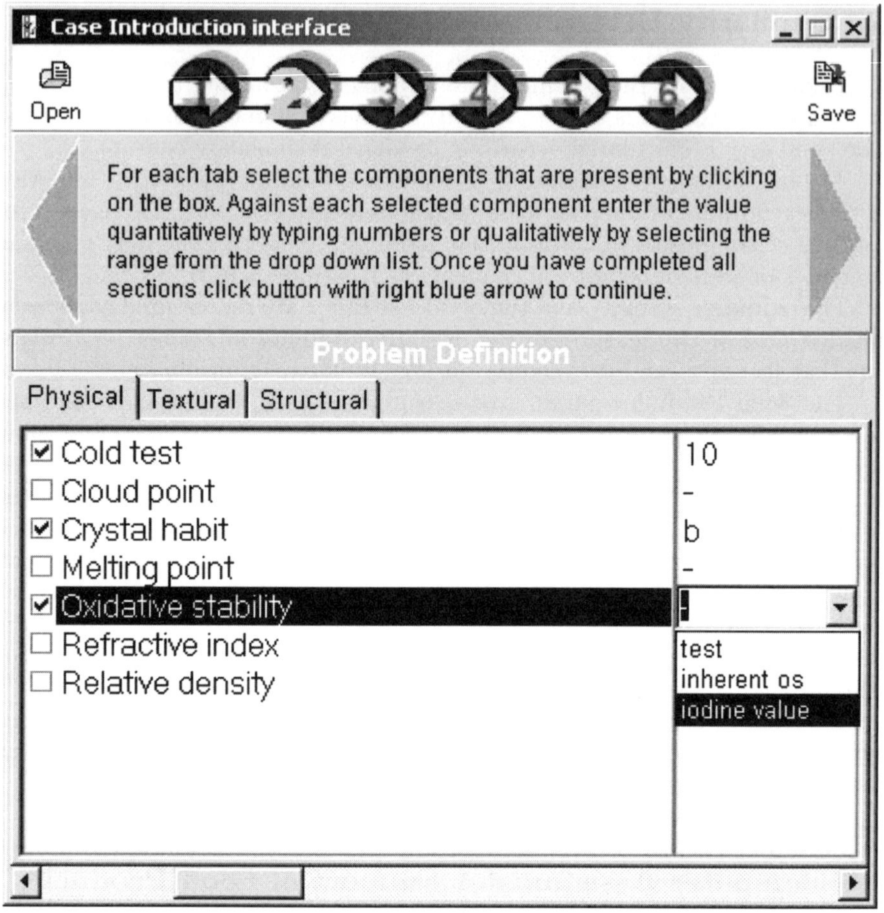

Fig. 6.2. The case definition form as an interface example of the computer assistant

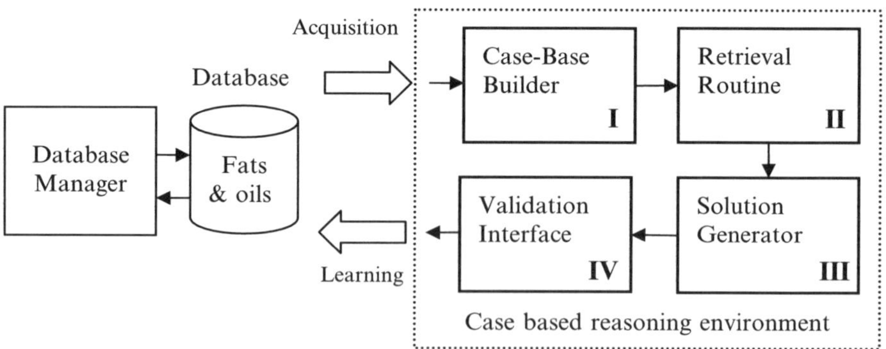

Fig. 6.3. The structure of design assistant for food product formulation

the data source. The current design problem as well as the structure of case representation is introduced by means of descriptive language before running of the tool. Using similarity measures implemented in the retrieval routine a set of similar past cases to the current design situation is retrieved – II. Based on the adaptation algorithm, prior solutions are used to propose a solution to the current situation – III. The proposed solution is then validated by the user – IV. If it succeeds, the working solution together with current problem is stored in a pre-defined format as a new information source.

The design support procedure can be described as follows. The functionality of the product in development converted to physical properties (after properties design) is used to identify the oils with most similar properties. The design assistant modifies the retrieved composition in the adaptation phase. By analyzing the modified composition during validation the necessary processing of fats and oils raw materials can be selected. The desired composition and physical properties can be achieved by blending, hydrogenation, fractionation, interesterification and emulsification. This is already a task of process design. Therefore, the design assistant supports the intermediate step of food ingredient formulation – product design that is considered after properties determination and before process selection and design.

6.6 Example: Cookie Filler Development

The goal is to design specific cookie filler with eating character and flavour stability, high plasticity, with soft and delicate mouth feel. The product must have high oxidative stability.

For a fat blend to be plastic, it must have both a solid and liquid phase. The ratio of these two phases determines its consistency. In addition, a melting point must be lower than body temperature for good eating characteristics. Thus, the solid fat indices curve must be steep around room temperature. The product requires as low an iodine value as possible for oxidative stability. Summarizing the requirements, the product features must satisfy to values presented in the Table 6.4. Crystal habit β' is desired for solidified product because it promotes good plasticity.

Table 6.4. The problem description of fats and oils product development

Feature	Value
Iodine value	20
AOM test (hours)	72
Melting point	25
Crystal habit	β'
Solid fat indices	(10°C, 60%), (20°C, 30%), (23°C, 10%), (25°C, 0%).

Table 6.5. The result composition of fats and oils product

Feature	Component	Value
Fatty acid content	Caprylic	8.5
	Capric	3.5
	Lauric	42.6
	Linoleic	2.8
	Myristic	23.6
	Oleic	4.9
	Palmitic	10.3
	Stearic	2.8
Triglyceride content	Trisaturated	79
	Disaturated	21

The case base according to proposed case structure was created from database of fats and oils blends. Palm kernel oil and coconut oil got the highest value of similarity for the introduced problem. Both oils have high result of Active Oxygen Method (AOM) – over 100 h. The fatty acid composition and triglyceride composition were adapted using general similarity concept for compositions part of case representation to get steeper SFI slope. The result is given in Table 6.5.

Analysing the modified composition during validation the necessary processing of fats and oils raw materials can be selected. The desired composition and physical properties can be achieved by the blending, hydrogenation, and emulsification of palm kernel and coconut oils.

7

Conceptual Design: Process Sequence Synthesis

The goal of conceptual design is to select the process operations and the interconnections among the units. The problem is difficult because very many process alternatives could be considered. There are many possibilities to consider with only a small chance to success. In some case it is possible to use design guidelines (like rules or heuristics) to make some decisions about the structure of the flowsheet and to set the values of some of the design variables. But in many cases, no heuristics are available and therefore direct reference to design experience can be used as a basis for making decisions.

In this chapter the conceptual design of wastewater treatment system is considered where the design experience in form of cases are combined with design heuristics which makes decision support more effective.

7.1 Introduction

The conceptual design of wastewater treatment systems is a demanding task for environmental engineers. Usually the task facing an engineer is to determine the levels of treatment that must be achieved and a sequence of methods that can be used to remove or to modify the constituents found in wastewater in order to reduce the environmental impact and to meet ecological requirements. The solution of this task requires detailed analyses of local conditions, needs and application of scientific knowledge and engineering judgment based on past experience.

The design of wastewater treatment system requires significant engineering experience, intuition and creativeness. Usually the task facing the engineer is to determine the levels of treatment that must be achieved and a sequence of technologies that is able to perform the necessary treatment in order to reduce environmental impact and meet ecological requirements. The solution of this task requires detailed analyses of local conditions, needs and application of scientific knowledge and engineering judgment based on past experience.

Y. Avramenko and A. Kraslawski: *Case-Based Design*, Studies in Computational Intelligence (SCI) **87**, 117–129 (2008)
www.springerlink.com

The approaches that have been used in the conceptual design of wastewater treatment processes include mathematical programming, thermodynamic methods and artificial intelligence. Such techniques for the decision support as inductive learning (Yang and Kao, 1996), hierarchical procedure (Freitas et al., 2000) and fuzzy sets (Krovvidy et al., 1994) applied to the design problem of wastewater treatment.

Most of these methods are based on complicated mathematical algorithms that generalize the experience of the designers. However, frequently creativeness and experience of the engineer is often difficult to be embedded in the algorithms. Meanwhile, in many cases the wastewater problems are similar and they can be solved in similar way.

Nowadays, with the increasing number of complex wastes that are the result of industrial operations, it is more than ever necessary for an engineer to review all the available methods, processes, systems and equipment in the light of demands and conditions, and to apply any single method or combination of them in the given situation. In addition, in many cases the wastewater problems are similar and they can be solved in a similar way. A design engineer may encounter a problem in determining the similar elements between a new problem and massive historical data; moreover, the similarities are often unnoticeable.

The developed methodology of case-based design support has been applied to assist environmental engineers in phase C (conceptual) of the process design of wastewater treatment systems.

The objective of the design phase is to construct sequence of processes for wastewater treatment that is able to treat a wastewater flow with given characteristics.

The task is to support the preliminary design of wastewater treatment system, to help the engineers to avoid a time-consuming selection of feasible and appropriate treatment technologies for the new wastes with maximum reuse of past design histories.

Using data about inlet water characteristics (such as flowrate, solids contents, biochemical oxygen demand etc.) and purity requirement for outlet water the sequence of appropriate treatment processes is determined (Fig. 7.1).

7.2 Case Base of Wastewater Treatment Systems

The case base collects the detailed description of treatment system including process sequence, function description, inlet and outlet water characteristics, technical features and also cost and energy consumption. It contains data acquired from the environmental engineers and wastewater treatment plants managers as well as other pertinent information taken from literature. The library of cases covers six sectors of industry producing wastewater:

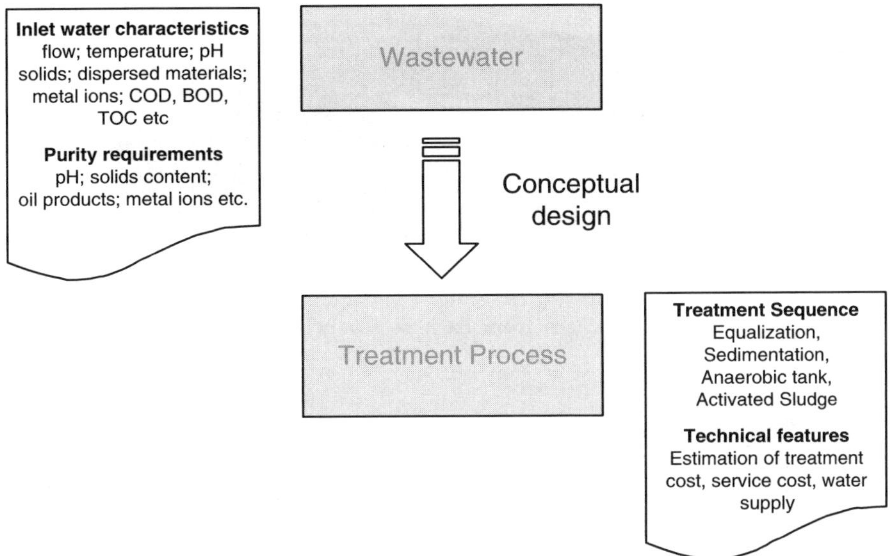

Fig. 7.1. Conceptual synthesis of wastewater treatment system

distillery, pulp and paper, metal finishing, textile, rubber and latex, and tannery. The industrial sectors are supplemented by municipal wastewater treatment plants. The scope of the case base is shown in Fig. 7.2.

However, the overall data collection includes also the technology descriptions which are used in the wastewater treatment schemas, the equipment database presenting wastewater treatment modules and aggregates, and methods knowledge base with the principles of construction of treatment sequences.

The technology database provides the user with a comprehensive overview of 20 processes used for wastewater treatment. They are grouped according to stages of wastewater treatment. The description of a technology includes not only basic principles but simple model, examples of applications and visualization illustrating process of treatment provided by the technology. The list of technologies included in the data base are presented in Fig. 7.3.

The equipment database contains data on individual treatment units manufactured by the different companies. It includes the name of unit, purpose, an operation description, effectiveness and limitation of application, a technical drawing, cost as well as a link to the company-producer database. It covers more than 200 units of equipment and represents 40 manufacturers.

The methods knowledge base includes rules of applications of methods for removal of harmful factors and efficiency for each factor, ranges, fields as well as the cost of application.

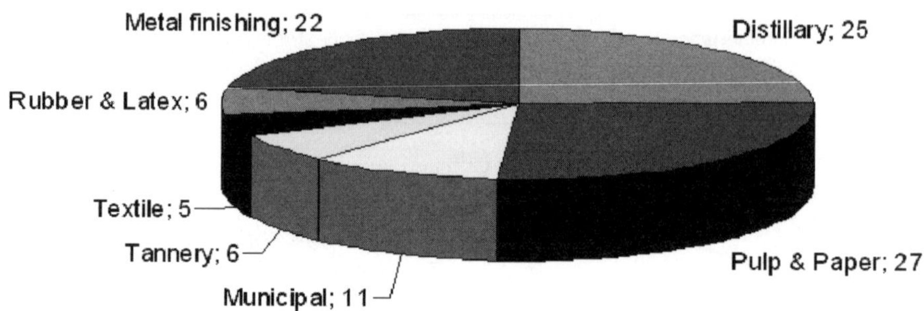

Fig. 7.2. The number of cases from each sector in the case base of wastewater treatment

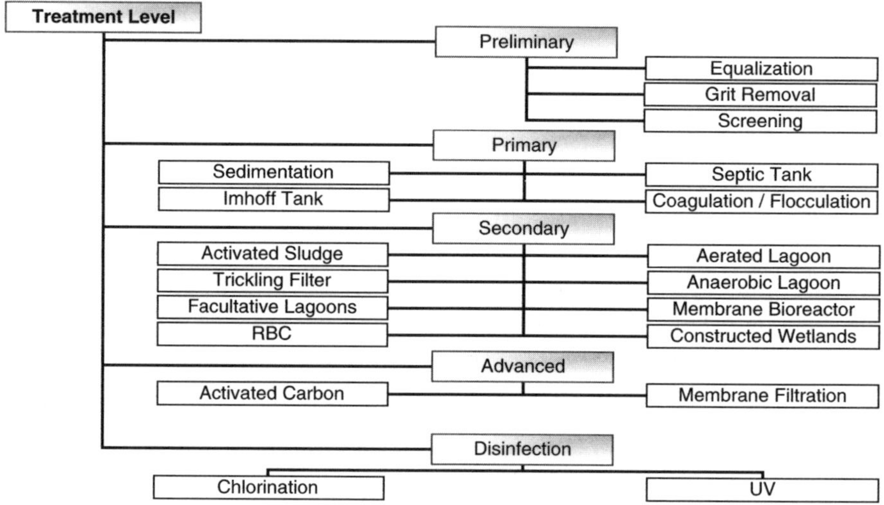

Fig. 7.3. The treatment method tree of the technology database

7.3 Case Representation for Wastewater Treatment Problems

The case of wastewater treatment is described as a set of features and their corresponding values. Some features are of complex type (i.e. they are represented in terms of other features).

The set of the essential features for the correct identification of the case has been divided into several parts (Fig. 7.4):

(1) Inlet water characteristics: such values as suspended solids contents, pH, biochemical oxygen demand (BOD), total organic carbon (TOC), alkalinity, heavy metals ions, coliform content etc.;
(2) Outlet water characteristics: data on water characteristics after treatment or treatment efficiency – similar to inlet parameters;

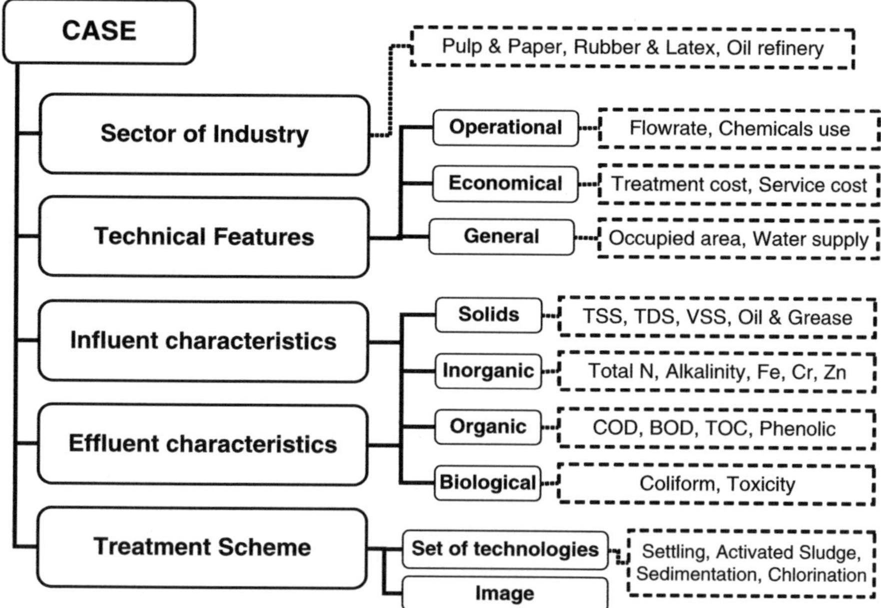

Fig. 7.4. Representation of a case of wastewater treatment (with examples)

(3) Technical features: performance, power consumption, occupied area, water circuit, treatment cost, etc.;
(4) Treatment: flowsheet representation, list of technologies for wastewater treatment.

Two groups of parameters are usually used as a problem description to identify an appropriate combination from the past experience. Other groups of characteristics are the solution part. However, such a division into problem and solution parts is not rigid. Some parameters from the third group can be considered as constraints, and therefore, can be included in the solution part to find a case that satisfies these constrains or approaches these parameters.

The case base collects detailed descriptions of existing treatment systems including treatment sequence, influent and effluent water characteristics, technical features and also water supply cost and energy and chemicals consumptions. It contains knowledge acquired from engineers as well as other pertinent information taken from literature. The case base includes case studies obtained from municipal and industrial wastewater treatment plants from Asia and Europe. The industrial sectors include pulp and paper mills, alcohol distilleries, tanneries, rubber and latex processing, textile and garment manufacturing and metal-finishing units. The weights of importance assigned to all features of the case structure based on expert opinions differ for each sector.

Depending on the industry sector the structure of cases varies. A certain set of wastewater characteristics corresponds to a sector of industry. Some characteristics may be grouped in on entity to provide more flexible retrieval. For example, for metal finishing problems the concentrations of heavy metals ions (Fe^{3+}/Fe^{2+}; Ni^{2+}; Cu^{2+}...) can be combined in one set because they can be treated in a similar way. An exact match is not necessary for such constituents but similar ions might be found. This is an example of a composite value: a set of structured elements. The difference value between ions might depend on their relative positions in the periodic table and ion charges. Thus, each ion is represented as an entity but of defined structure including its concentration in the wastewater flow. Another example is salt content. The characteristic can be represented by anions concentrations, which mostly can be combined in one feature summarizing anions.

7.4 Computer Assistant for Wastewater Treatment Synthesis

The computer assistant is a decision supporting system. The system is organized in conventional way: it has separate elements for the passive, data part and the active, program part. The data part includes the base of past cases of wastewater treatment and the database of technologies applied to wastewater treatment. The active part of the system is composed of four components: Database Manager (also called Reference Library), Treatment Sequence Builder, Case-Based Reasoner, which communicates with the user by means of the System Interface. The overall structure of the system is illustrated in Fig. 7.5.

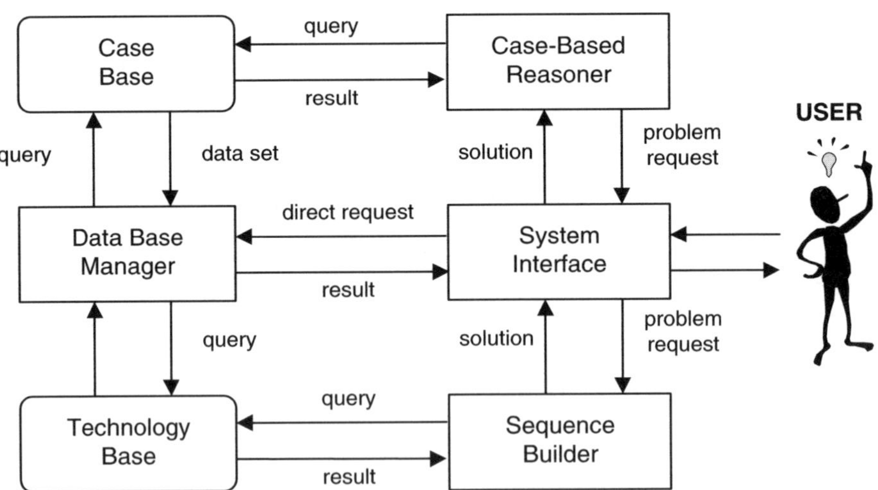

Fig. 7.5. The structure of DSS for wastewater treatment synthesis

The Database Manager is responsible for the resource management and the consistency of the database. It performs any search operations in the database space, and constructs the report as a result to the user's query.

The individual treatment technologies are usually classified as physical operations, chemical and biological processes. But in the technology base of the system, the unit operations and processes are grouped according to the level of the provided treatment. There are preliminary, primary, secondary, and advanced treatment technologies presented in the library. Each group can be expanded into subgroups. For example, the secondary treatment group (biological treatment) is divided into aerobic and anaerobic suspended growth methods, aerobic and anaerobic attached growth processes, and lagoon methods. Further, a subgroup is splitted into types of unit processes.

The databases of the system are capable of use as stand-alone tools.

The user introduces his/her problem through input forms in the system interface. The Case-Based Reasoner tries to find the most similar problem and applies an old solution of the retrieved similar problem to an actual situation. If the proper solution is not found, the treatment sequence can be built by a special algorithm realized in the Sequence Builder.

7.4.1 Database Manager – Reference Library

The Database Manager is able to navigate on overall data collections, i.e. treatment system case base, technology base, equipment database and treatment methods base. However, the interface is specially designed to work with the technology base. The manger over technology base is called the reference library.

The purpose of the reference library (RL) is to provide the user with the comprehensive overview of processes and operations used for water treatment through visualization of real-life units. The general description of the wastewater treatment technology is supplemented by the theoretical background as well as a worked out example and an Excel spreadsheet model. The user can modify the selected parameters in the spreadsheet to understand their effect on the unit performance. To illustrate the principles and the units used in the water treatment, the system contains the schematic pictures, photographs, 3D images and animation of the corresponding operations. The animation provides the basic understanding of how the process is realized.

By using the convenient navigation tools located in the left part of the workspace of the RL, supplied with the additional facilities (alphabetic indexing, text searching), it is possible to get access to knowledge and data stored in RL. The description is given in the right part of the library's workspace. At the top, there is an introduction to the selected technology. At the bottom the details of technology are given (Fig. 7.6).

The particular treatment processes are usually classified as physical operations, chemical and biological processes. Reference Library supports several

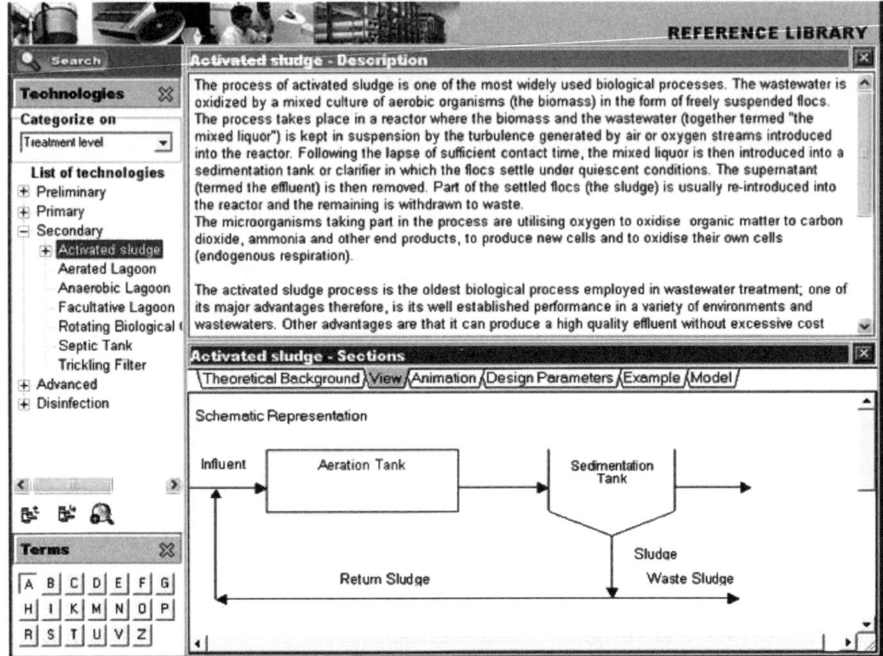

Fig. 7.6. The layout of the Reference Library

classifications of the unit operations and processes. They are grouped according to the level of the provided treatment (preliminary, primary, secondary, and advanced treatment), type of unit operations (physical, chemical, biological) and in the alphabetic order.

The group of primary treatment contains mostly physical operation, such as screening, sedimentations, flotation. The secondary treatment group is represented exclusively by biological processes. Each group can be expanded into subgroups. For example, the secondary treatment group (biological treatment) is divided into aerobic and anaerobic suspended growth methods, aerobic and anaerobic attached growth processes, and lagoon methods. Further, a subgroup is splitted into types of unit processes.

The Reference Library is supplemented with a glossary section, where the user can find definitions of terms used in the text of the RL concerning water and wastewater parameters and treatment processes.

7.4.2 Case-Based Reasoner

The Case-Based Reasoner (also Case Study Manager – CM) accumulates the specific design experience contained in real life situations, and tries to reuse it when solving new user's problems. The manager performs the retrieval of the most similar cases to the current problem from the case base containing the past situations of wastewater treatment. It utilises the case-based reasoning approach in solving new design task. There has been developed the following method in order to define a similarity between the cases containing both numeric and textual-symbolic information.

The case base of the manager includes case studies obtained from municipal and industrial wastewater treatment plants from both Asia and Europe. The industrial sectors include pulp and paper mills, alcohol distilleries, tanneries, rubber and latex processing, textile and garment manufacturing and metal-finishing units.

The representation of the case includes description of influent and effluent water characteristics, type of industry, description of used technology, and technical parameters of treatment operations such as flow rate, cost of treatment, water supply etc.

The CM is organized in similar way to the RL (see Fig. 7.7). The left part in the workspace is used for navigation in the case base. It also includes searching facility and tool for finding a set of the most similar cases. Once relevant cases have been retrieved from the case base, the user can browse through them in order to select the most applicable ones for the current situation. The right part of the workspace contains the general description of a case (at the top) and the case details (at the bottom). The characteristics of the wastewater are grouped into the following sub-classes: physical, inorganic, organic and biological.

A new problem is introduced via convenient input form (see Fig. 7.8).

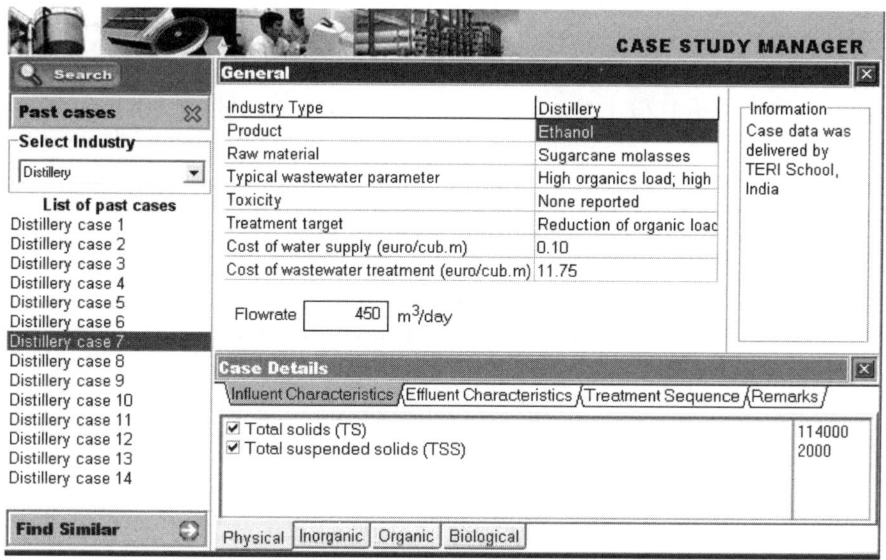

Fig. 7.7. The layout of the Case Study Manager

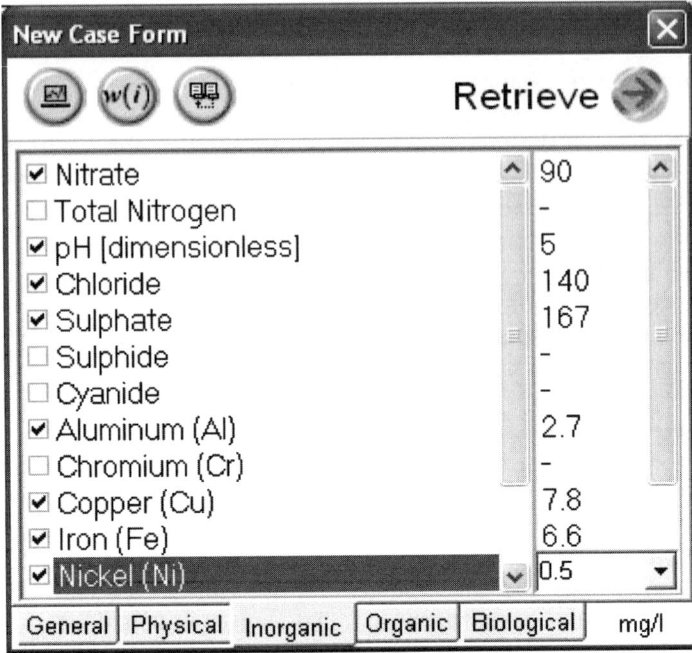

Fig. 7.8. Introduction of a new problem

7.4.3 Treatment Builder

Treatment builder is able to construct the treatment sequence for wastewater with specific characteristics based on basic principles and the heuristics. It supplements the Case-Based Reasoner in decision support of design of wastewater treatment systems. The builder has two components: treatment adviser (TA) and process builder (PB).

The TA generates a simple sequence of treatment technologies for a given water characteristics. It analyses the influent water characteristics and selects, performing the original algorithm based on set theory, the methods of treatment. The algorithm of selection is based on the search among the water parameters, so-called harmful factors that have to be eliminated. The factors are determined by specific set of wastewater characteristics. For example, the phenol concentration in water above $50\,\mathrm{mg\,l^{-1}}$ and up to $500\,\mathrm{mg\,l^{-1}}$ defines the harmful factor "Middle concentrated phenol". Each harmful factor can be treated by a number of wastewater treatment technologies that are capable to remove the factor from wastewater. The stream may contain a number of harmful factors that can be processed by many sets of treatment methods. As a result of analysis, one or several treatment sequences are generated and then evaluated by economical and treatment efficiency criteria. The economic and efficiency evaluations are done based on data from the past applications of a technology.

The TA has the same design that the previous components of the education environmental. On the left side of the workspace there are the elements for navigation in the advisor's knowledge base, whereas the right part is for description of the classes of harmful factors and the list of technologies for their removal.

The process builder has the ability to construct the treatment sequence from the blocks. The user can drag and drop the blocks located on the screen. Each of the blocks represents a type of the treatment processes or specific part of the process. Blocks can be linked according to internal restrictions, rules and locations of connection points. When two or more blocks have been connected, flow animation or process visualization occurs. The interconnection rules integrated in the PB are based on a valid sequence matrix allowing the user to view when a treatment scheme is not feasible.

The constructed sequences can be saved as the file and restored in the next session. The main purpose of this component is to display a treatment sequence generated by the TA.

The design of the PB is not similar to others components of the presented DSS. The top of the workspace contains the small icons of the process divided to the sections of treatment levels. The icons can be dragged and placed to the white sheet of the workspace to be converted to process blocks. The constructed sequence can then be easily edited by removing of the existing block and adding new ones.

The treatment builder can also search for specific equipment or modules that are realized operations presenting in the selected sequences. In this it cooperate with the Database Manager by queries.

7.5 Example: Zinc-Plating Workshop Wastewater

The decision supporting system has been used for preliminary phase of conceptual design of treatment of a wastewater stream from an electroplating plant.

The wastewater flow of $5.5\,m^3\,h^{-1}$ of zinc-plating workshop is described by set of characteristics shown in the Table 7.1. There were selected four entities in the stream: E_1 – metals $\{Fe^{3+}/Fe^{2+};\ Ni^{2+};\ Zn^{2+};\ Cu^{2+};\ Al^{3+}\}$, E_2 – salts (phosphates, chlorides, sulphates, nitrates), E_3 – bionondegradable organics, and E_4 – pH of water. Each entity is represented by a complex data structure combining two or more data types. For example, E_1 is a set of subsets (group of metals with common properties creates a separate subset) and vectors, E_2 – a set of features, numeric values and hierarchical type, as there is a need to determine the similarity between chemical organic compounds based on their chemical structure (implemented as a hierarchical tree). The case base was restructured according to above representation of the case. Only the structurally similar cases are considered in the subsequent calculations. The weights of importance have a high priority for heavy metal concentration and salt content as the additional condition assigned to return water to the

Table 7.1. Problem input of example case

Parameter	Input (mg l^{-1})	Output (mg l^{-1})
pH	5.0	7.0
SO_4^{2-}	167	100
Cl^-	140	300
NO_3^-	90	40
Zn^{2+}	27.3	0.01
Ni^{2+}	0.5	0.01
Al^{3+}	2.7	0.04
Fe^{3+}/Fe^{2+}	6.6	0.01
Cu^{2+}	7.8	0.01
Formalin	0.3	0.2
Surface-active materials	1.8	0.1

Table 7.2. Description of selected case

Parameter	Input (mg l^{-1})	Output (mg l^{-1})
pH	4.0	8.0
Sulphates	~ 200	100
Chloride	~ 100	300
NO_3^-	60	40
Zn^{2+}	50–100	0.01
Cu^{2+}	50–100	0.01
Cr^{3+}	25–75	0.04
Fe^{3+}/Fe^{2+}	10–30	0.005
Cu^{2+}	50–100	0.01
Formalin	0.3	0.2
Surface-active materials	10	0.1

industrial process. The system selected a case described in the Table 7.2. The achieved similarity is around 78%. As it can be seen from the Table 7.2, the salt content is still high and the treatment flowsheet has to be adapted. Using the treatment builder component there were suggested several methods for reducing the salt content – electrodialysis, ion exchange, reverse osmosis, evaporation. After performing the economical analysis efficiency estimation, it resulted that the most suitable was ion-exchange method. The flowsheet of the most similar case and its additional part obtained from adaptation are shown in Fig. 7.9.

The described DSS can give answer to query what king of sequences of treatment operations should be used to process of a wastewater stream with certain characteristics. But in additional to such expert functionality, the system provides an opportunity for the users to learn wastewater treatment technologies and approaches to water conservation in several countries in Asia

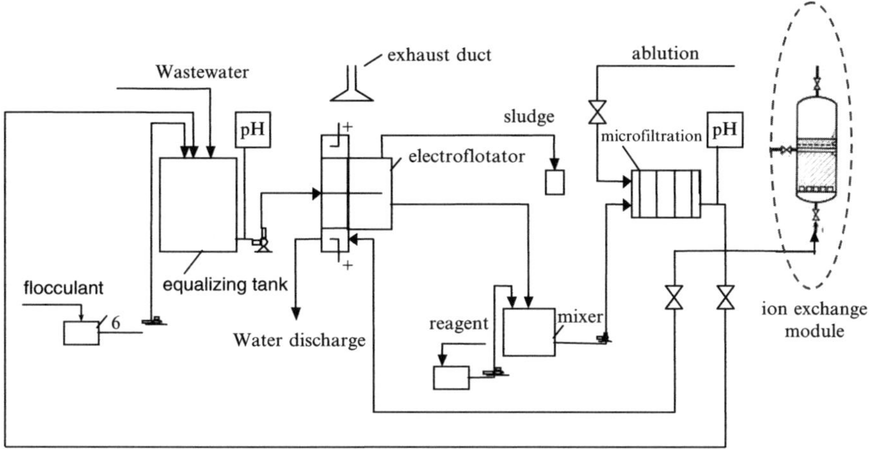

Fig. 7.9. Selected flowsheet and adapted part with ion exchange method

and Europe. They also can have an overview of the theory and practice of wastewater treatment technologies applicable to municipal and various industrial sectors in Asia and Europe. In such situation it could be regarded as a comprehensive decision supporting system for wastewater treatment.

8

Pre-Detailed Design: Process Model Selection

The chapter focuses on such design activity as the selection of proper model describing the processes and phenomena. This activity could take place as early as on Conceptual design phase to evaluate the alternatives and also on Detailed design to get more specific characteristics of designing process. Therefore, the applications of Case-based design supporting method to model selection process are united under pre-detailed design activity.

The procedure of model selection could not actually be a design supporting method itself. It is only initial step in further modelling and evaluation. But the design engineers in many cases meet serious difficulties in determination proper model type, acceptable simplifications of model, and initial state for the modelling. The CBD approach seems to be a good opportunity to overcome the difficulties and in such way to support the design process.

Two different applications of CBD for the model selection are considered in the chapter: selection the mixed integer nonlinear programming model for synthesis of distillation systems and selection of model for selective catalytic reduction of NOx with ammonia in forced unsteady-state reactors.

8.1 Introduction

The design and simulation of chemical reactors or distillation systems involves complex mathematical modelling containing the description of transport phenomena and reactions in processes that take place in multi-phase systems. The detailed mathematical models of these systems that account for inter and intra-phase gradients as well as the exchange or interaction between the phases are very complex and may be expressed in the form of several partial differential equations in two or three spatial coordinates and in time. In addition, these detailed models contain a large number of physical–chemical parameters. In case of unsteady-state reactor, there are regions of the parameter space in which the rates of some of the transport or reaction processes are much slower or faster as compared to the other parts of this space. It results in

Y. Avramenko and A. Kraslawski: *Case-Based Design*, Studies in Computational Intelligence (SCI) **87**, 131–152 (2008)
www.springerlink.com

a big complexity of the phenomena that take place inside the reactors making the models very often inadequate for analysis or computation. The numerical procedures to solve mathematical models related to forced unsteady state reactors required a lot of work focused on the reduction of computational time. This is realized by using highly-efficient numerical methods by simplifying mathematical models to one dimension and by fixing the boundary values in time at the beginning and at the end of the half cycle to avoid the need to solve over the long transient period before the establishment of the stationary conditions (Botar-Jid, 2007).

Despite a large body of literature in this field there is no standard procedure developed for the design of reverse flow reactors. Usually the design is carried out by trial and error coupled with extensive and tedious detailed numerical simulations. Similar situation takes place in design of complex distillation systems.

One of the distillation system synthesis methods is mixed integer nonlinear programming (MINLP). MINLP affords the possibility to execute the synthesis and system optimization simultaneously (Duran and Grossmann, 1986). The method has three steps: build a superstructure; generate the MINLP model of the superstructure; find the optimal structure and operation.

There are two main difficulties when using MINLP: generating an accurate MINLP model is a complicated task, and MINLP algorithms provide a global optimum, in the case of convex searching space. In regard to generating an accurate MINLP model, usually, related papers report a new MINLP model and superstructure, according to the problem under consideration, but the development of all of these superstructures requires considerable engineering experience. Up to know, there are only a few automatic combinatorial methods available for generation of the superstructure (Farkas et al., 2006). In regard to the MINLP algorithms, the distillation column design models include strongly non-convex functions; therefore, finding a global optimum is not ensured. In such cases, the result is dependent on the initial point of calculations.

Case-based design support method can be applied for finding a proper MINLP model with the superstructure and suggesting an initial point for performing design and optimization of a distillation system. After optimization of the selected MINLP model a solution of the corresponding distillation synthesis problem can be obtained.

The problem is stated as follows: Given an ideal or close to ideal mixture of arbitrary components is to be separated into a number of products of specified compositions by means of distillation. The objective is to get the proper model with the starting point for the process of synthesis of the distillation column or distillation sequence. The superstructure must include an initial structure for the design optimization. The process of design of considered stage is shown in Fig. 8.1.

To simplify the study of the applicability of CBR, only ideal mixture separation cases are considered.

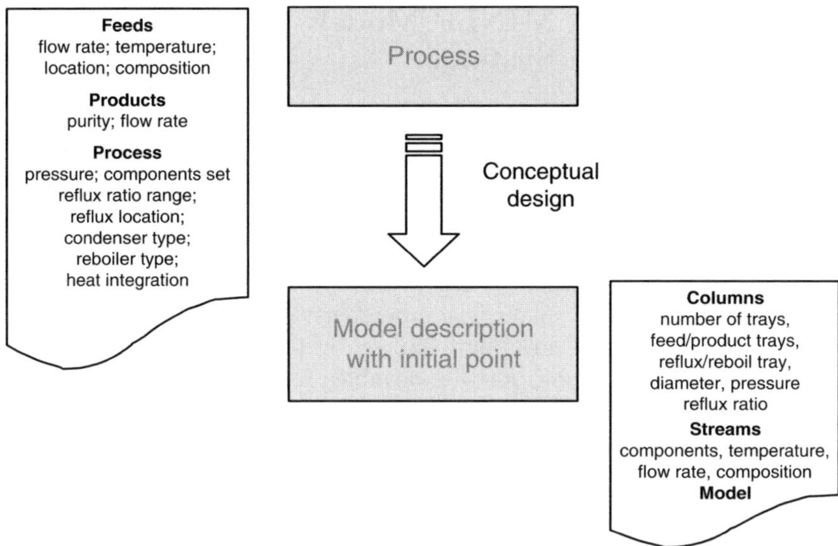

Fig. 8.1. Selection of MINLP model for synthesis of distillation systems

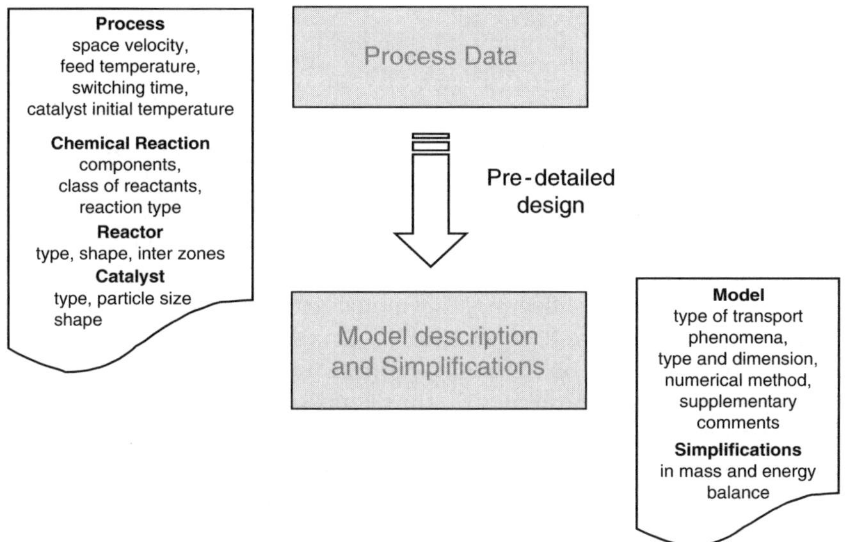

Fig. 8.2. Selection of reverse flow reactor model for forced unsteady-state problem

The case-based model selection is also used to provide a model for reverse flow reactor supplemented with necessary computational information for the modelling. The overall problem for forced unsteady-state reactor model selection is shown in Fig. 8.2.

8.2 Case Library of MINLP Model and Distillation Systems

The library of cases is built based on the detailed distillation examples with reproducible MINLP models that have been published in other papers. The case library contains 26 cases of separation of ideal mixtures for up to five components. The descriptions of the stored cases are given in Table 8.1.

The case library includes only cases with reproducible MINLP models. The representation of a model involves a superstructure, the set of variables and parameters, the mass and enthalpy balances, and other constraints. However, usually only the superstructure, the variables, and the main equations are detailed in the source articles; e.g. the equilibrium models and the basic mass balances are not represented. The articles contain the hints and notes, which can be helpful in regard to using a model. To provide the instructions for using the MINLP model, the original articles have been included in the case library as PDF files.

8.3 Representation of Models for Synthesis of Distillation Systems

Cases are represented as entities with the set of features, their values and relations. The model, consisting of a superstructure, a set of variables and parameters, the mass and enthalpy balances and other constraints, together with the flowsheet and its mathematical representation builds the solutions part of a case. The form of model varies with different layouts of distillation system. Therefore, the structure of entities involving the model specifications is dependent on certain characteristics of the distillation system.

Three characteristics affecting the model structure have been selected. The first one is the type of heat integration. A system can be either with heat integration or without it or thermally coupled. In a single column configuration only a non-heat integrated structure is possible. Single column configurations and models do not include the mass balances for the connections of distillation columns; and so, these models cannot be used for problems with three or more products. In addition, there is the dissimilarity between models with single and multiple feeds. Therefore the number of products (with only two grades: two or more products) and number of feeds (also only two grades: one or more) affect the structure of the entities as well. The total number of different structures being determined by a combination of these characteristics is 12 ($3 \times 2 \times 2 = 12$). But some combinations are not realistic. The different structures presented in the case library are given in Table 8.2.

Table 8.1. Stored cases in the library

	Mixture	Sharp Separation	Heat Integration	Reference
1	Propane; *iso*-butane; *n*-butane	No	No	Example 1 of Aggarwal and Floudas, 1992
2	Propane; *iso*-butane; *n*-butane	No	Yes	Example 1 of Aggarwal and Floudas, 1992
3	*n*-Butane; *n*-pentane; *n*-hexane; *n*-heptane	No	Yes	Example 2 of Aggarwal and Floudas, 1992
4	Benzene; toluene; *o*-xylene	Yes	No	Example MF1 of Viswanathan and Grossmann, 1993b
5	*n*-Hexane; *n*-heptane; *n*-nonane	Yes	No	Example MF2 of Viswanathan and Grossmann, 1993b
6	Acetone; acetonitrile; water	Yes	No	Example MF3 of Viswanathan and Grossmann, 1993b
7	Methanol; water	Yes	No	Example MF5 of Viswanathan and Grossmann, 1993b
8	Benzene; toluene; *o*-xylene	Yes	No	Example Ternary 1 of Viswanathan and Grossmann, 1993a
9	Benzene; toluene; *o*-xylene	Yes	No	Example Ternary 2 of Viswanathan and Grossmann, 1993a
10	Acetone; acetonitrile; water	Yes	No	Example Unit of Viswanathan and Grossmann, 1993a
11	Benzene; toluene; *o*-xylene; diphenyl	Yes	No	Example 1 of Novak et al., 1996
12	Benzene; toluene; *o*-xylene; diphenyl	Yes	Yes	Example 2 of Novak et al., 1996
13	Propane; butane; pentane; hexane	Yes	No	Example 1 of Yeomans and Grossmann, 1999
14	Propane; butane; pentane; hexane	Yes	Yes	Example 1 of Yeomans and Grossmann, 1999
15	Propane; butane; pentane; hexane	Yes	No	Example 1 of Yeomans and Grossmann, 1999
16	Propane; *n*-butane; *n*-pentane; *n*-hexane	Yes	No	Example 2 of Caballero and Grossmann, 1999
17	Propane; *n*-butane; *n*-pentane; *n*-hexane	Yes	Yes	Example 3 of Caballero and Grossmann, 1999

(continued)

Table 8.1. (Continued)

	Mixture	Sharp Separation	Heat Integration	Reference
18	Methylacetylene; propane; *n*-butane; *n*-pentane; *n*-hexane	Yes	No	Example 4 of Caballero and Grossmann, 1999
19	Methylacetylene; propane; *n*-butane; *n*-pentane; *n*-hexane	Yes	Yes	Example 5 of Caballero and Grossmann, 1999
20	Benzene; toluene	Yes	No	Example 1 of Yeomans and Grossmann, 2000
21	Benzene; toluene	Yes	No	Example 3 of Yeomans and Grossmann, 2000
22	*n*-Butane; *n*-pentane; *n*-hexane	Yes	No	Example 4 of Yeomans and Grossmann, 2000
23	Benzene; toluene; *o*-xylene	Yes	No	Example 5 of Yeomans and Grossmann, 2000
24	*n*-Pentane; *n*-hexane; *n*-heptane	Yes	Thermally linked	Example 5.1 of Yeomans and Grossmann, 2000
25	Benzene; toluene; *o*-xylene	Yes	Thermally linked	Example 1 of Caballero and Grossmann, 2001
26	*n*-Pentane; *n*-hexane; *n*-heptane; *n*-octane; *n*-nonane	Yes	Thermally linked	Example 2 of Caballero and Grossmann, 2001

Table 8.2. The list of different structures

Name	Heat integration	Number of feeds	Number of products
S_1	No	1	2
S_2	No	1	More than 2
S_3	No	More than 1	2
S_4	No	More than 1	More than 2
S_5	Normal	1	2 or more
S_6	Thermally coupled	1	More than 2

The entities with different structures are structurally dissimilar and they are not considered together during the retrieval procedure. Only entities which belong to the same type ($E_1 = <S_a, V_c>$, $E_2 = <S_b, V_d>$, $a = b$) are compared.

The problem part of the cases contains the same features for all structures (except those listed above). The list of features describing the problem is shown in Table 8.3.

Table 8.3. The list of features of problem description

Feature	Type
Components	Set of elements
Feeds	Set of elements
Products	Set of elements
Sharp separation	Logical
Maximum number of trays per column	Numeric
Additional constrains	Textual

Table 8.4. The structure of entities

Entity 'Feed'		*Entity 'Product'*		*Entity 'Component'*	
Feature	Type	Feature	Type	Feature	Type
Flow rate	Numeric	Type	Hierarchal	Flow rate	Numeric
Temperature	Numeric	Molar weight	Numeric	Composition	Vector
Pressure	Numeric	Boling point	Numeric		
Composition	Vector				

The advantage of the described concept consists in the state that each feature can be represented as a new entity. Each element of a set in the list of features 'Components', 'Feeds' and 'Products' is an entity with its own structure. Hence, these features are the features with composite values – sets of structured elements. The lists of features of those entities are given in Table 8.4.

The feature 'Type' in the entity 'Component' represents a chemical nature of a component. The 'type' of a component is based on its chemical structure and represented as a dangling node of the hierarchy of groups of chemical compounds. It is an example of a hierarchy with assigned values. A corresponding similarity tree of widely spread chemical compounds in respect to distillation problems has been constructed.

The cases in the case library are previously published distillation problems with reproducible MINLP models published in scientific papers. Each case contains a problem description and the mathematical representation of its solution.

The solution of a similar problem is given as an initial in optimization to increase greatly the probability to find the global optimum. The articles report usually a flowsheet supplemented with a dataset as a solution for a problem.

A flowsheet is represented as a graph. An example of graph representation of the flowsheet (Yeomans and Grossmann, 1999) is shown in Fig. 8.3. In this graph the nodes are the feed (F1), the distillation columns (C1, C2, C3), the heat-exchangers (condensers: Con1,...; and reboilers: Reb1,...), the mixers/splitters (MS1, MS2,...) and the products (P1, P2,...); the edges are

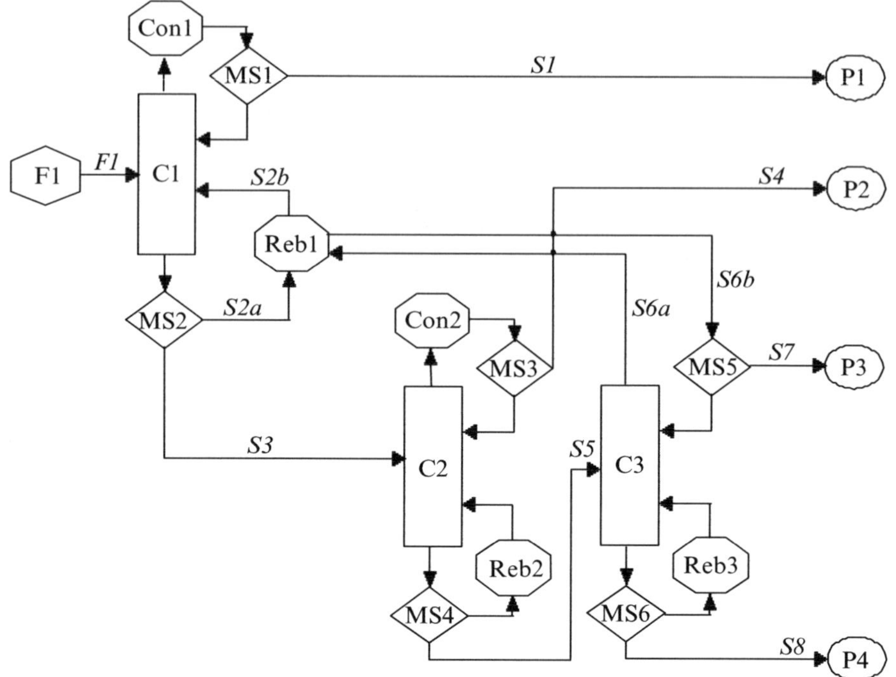

Fig. 8.3. Graph representation of flowsheet

the flows between the units. This graph can be represented in matrix form (node–node matrix). In this matrix $a_{ij} = 1$ if there is connection from node i to node j, otherwise, $a_{ij} = 0$.

Many flows are supplemented with attributes such as temperature, flow rate, composition. Such flows have the captions (e.g. S1, S6b) in the graph. These flows are represented in separate edge–node matrix, which contains the starting and ending nodes of the flows.

In the graph representation (Fig. 8.3) only simple columns are used, with maximum three inputs and two outputs. In case of thermally coupled flowsheets a possible rearrangement of the complex columns is used. If two flows between two columns have reverse direction then these flows pair is called 'thermally coupled'. The thermally coupled complex columns are represented as composed of two parts–upper and lower separate columns.

The solution is represented by the graph, the node–node matrix and the edge–node matrix as well as the detailed data of units and flows, such as:

Distillation columns	– Number of trays
	– Diameter (m)
	– Input/output trays
	– Pressure (bar)
	– Reflux ratio
Heat exchangers	– Area (m^2)
	– Heat flowrate (MW)
	– Utility
flows	– Temperature (K)
	– Flowrate (kmol h^{-1})
	– Set of components
	– Mole fraction of components

In case of heat integrated columns the flows go through heat exchangers. Heat exchanger changes the temperature and physical condition of the flow. However, the rate of temperature changing is unknown. Therefore, these flows are marked with the same number, and distinguished with small letters (e.g. S2a, S2b), but only the data of the flow before heat exchanger are reported.

8.4 Similarity Calculation for Distillation Problems

During case retrieval only the cases being represented by one structure are considered together. The cases that have different case structure (see Table 8.2) are regarded as dissimilar.

Similarity between cases with one structure is determined according global similarity formula described in the previous chapter. But feature could be of complex structure and the determination of similarity between them could differ.

The similarity between component sets is very important and must be applied first. It must be determined which component in the source case corresponds to a certain component in the target case. In the simplest case, the sets of components of the target case and the source case are identical. Otherwise, the most similar sequence of components must be determined, and identical components often do not create the corresponding pairs. For instance, the components set of the target case (according to Yeomans and Grossmann, 2000) is *n*-butane, *n*-pentane, and *n*-hexane. The components set of the source case is *n*-pentane, *n*-hexane, and nheptane. The *n*-pentane and *n*-hexane components are present in both cases, and it is evident to assign them to each other in the target case and in the source cases. The third pair of the components then is *n*-butane (the target case) and *n*-heptane (the source case). However, there is a problem with this assignment, because of the fact that *n*-butane in the target case is the most volatile component, whereas

n-heptane, the pair of n-butane in the source case, is the less-volatile component. Thus, the solution of the source case cannot be used for the solution of the target case.

To overcome these difficulties, during the matching of the components, the primary assumption is the volatility order of the components, and the secondary assumption is the nature of the components. The component pairs in the previous example are n-butane–n-pentane, n-pentane–n-hexane, and n-hexanen–heptane. In this case, the solution of the source case can be used to solve the target case.

To calculate the similarity, five attributes are used: components, boiling points of components, molar masses of components, feed, and product composition (mole fraction).

The similarity of components is based on their chemical structure. The similarity tree, which includes all components in the case library (Fig. 8.4), has been built. In the similarity tree, the nodes represent the basic groups of chemical components. To each component group, a numeric similarity value was assigned. The similarity value of two components is the value of the nearest common node in the tree. For example, when comparing n-butane and methanol, the nearest common node is the "organic" node; therefore, the similarity value is 0.2. The more similar the components, the greater the similarity value between them. For identical components, the similarity value is 1. It may happen that cases with different numbers of products are compared. In such cases, there are components in one set that have no corresponding components in another set. For these matchless components, the nearest common node is the "components" node; therefore, the similarity value is 0.

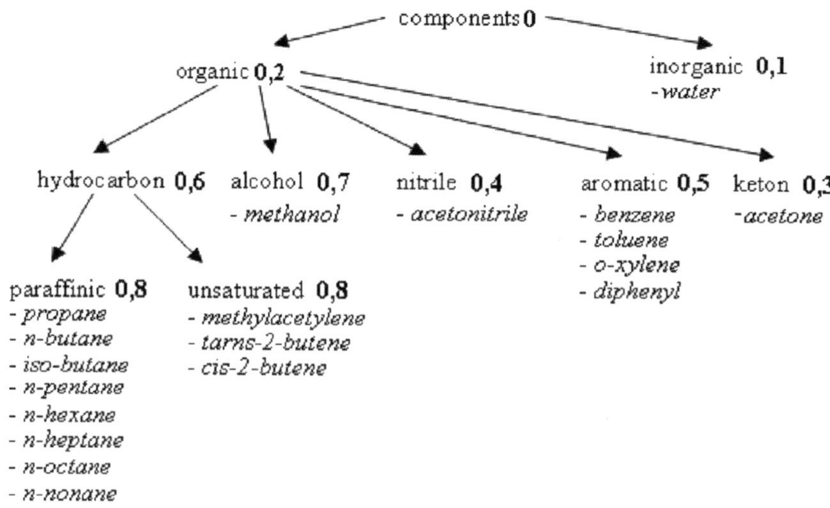

Fig. 8.4. Similarity tree of components

Because only problems that contain ideal mixtures are stored in the case library, the comparison of components, based on chemical structure of the components, is suitable.

Feed and product composition are compared using difference formula for vectors.

Three most-similar cases are selected as potential solutions, and, according to the actual requirements and engineering experiences, the most useful model is chosen. Because of the complexity of the distillation problems, there is no automatic adaptation of the found solution. The task of the designer is the modification of the MINLP model and the reuse of the solution of the chosen case as an initial point for design and optimization.

8.5 Computer Tool for Selection of MINLP Model

The computer tool is a complex of modules where case-based reasoning is implemented with the CaBaReEn environment (Fig. 8.5). The database containing descriptions of distillation problems and reproducible MINLP models are managed by separate tool. The representation of the solution (MINLP problems) requires the functionality to store and describe the flowsheet by means of graph, matrix, and schematic image. The structure of the problem description is flexible and requires changeable representation. The corresponding editors for the solution and problem parts have been designed (see Fig. 8.6).

The case base is created from the database according to the given case structure (see Table 8.3) by a module of the environment – Case-Base Compiler.

The new problem is introduced using a corresponding interface – Problem Director. The modification of the problem structure can be performed in the director. Using the Similarity Measurer module, which can vary similarity functions, a set of similar problems to model description is retrieved.

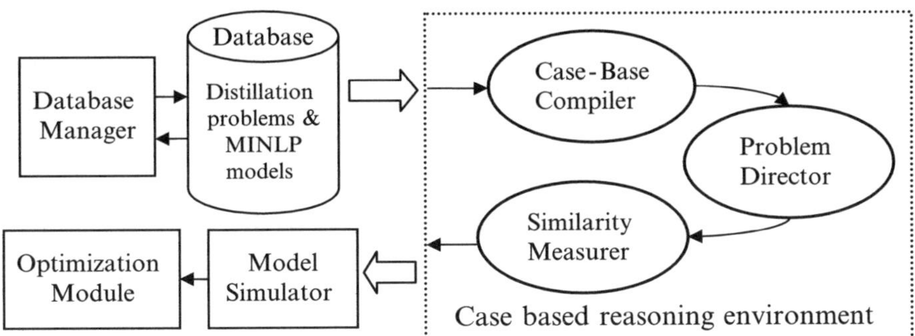

Fig. 8.5. Structure of the CBR system for support of distillation system synthesis

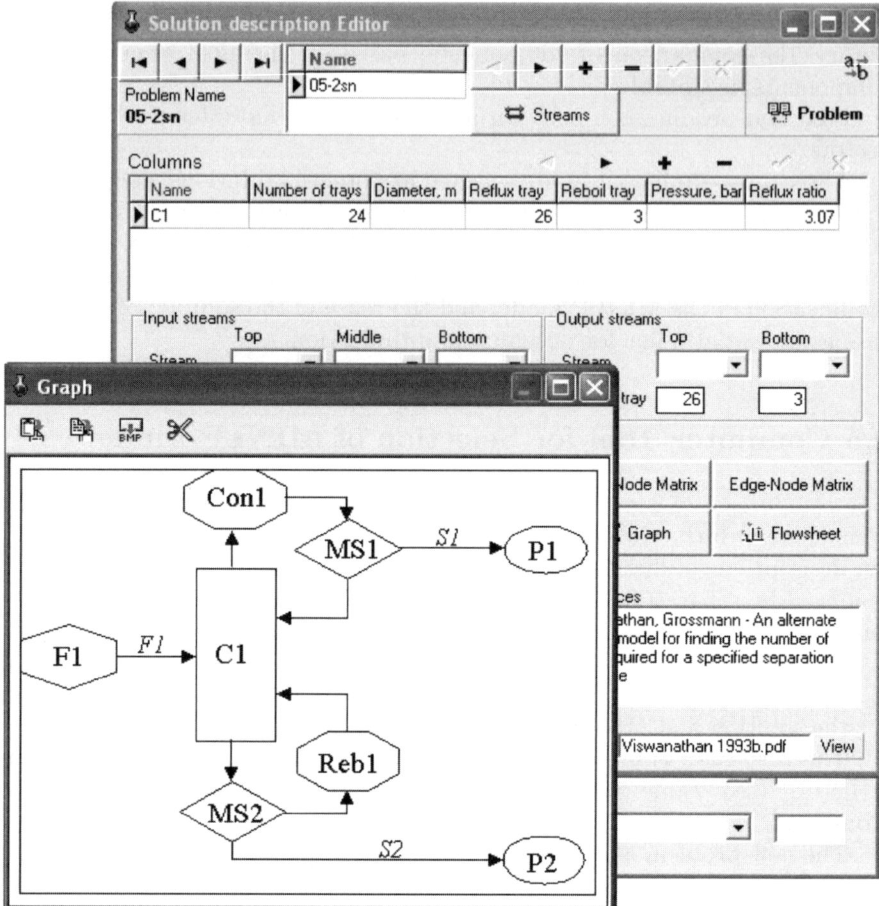

Fig. 8.6. A screen of a solution description of the CBR system for model selection for distillation system synthesis

The retrieved MINLP model and the superstructure as the initial point are sent to the simulator. The model is solved and then optimized under the conditions and parameters of the current distillation problem.

8.6 Example: Separation of Heptane–Toluene Mixture

There is given a heptane–toluene mixture. The flowrate of the equimolar (0.5, 0.5) feed is 100 kmol h^{-1}. The target is to separate the mixture into pure components with 95% purity requirement at the top and at the bottom.

It is a sharp separation problem and a single-column configuration should be used, which means that the searched structure is not heat-integrated. There

Table 8.5. Retrieval results for heptane–toluene mixture problem

	Source Case 1	Source Case 2	Source Case 3	Source Case 4
Problem Published Originally	Example Ternary 1	Example Ternary 2	Example Unit	Example 1
System	Benzene	Benzene	Acetone	Benzene
	Toluene o-Xylene	Toluene o-Xylene	Acetonitrile Water	Toluene
sim_c	0.400	0.400	0.133	0.600
sim_t	0.777	0.777	0.767	0.967
sim_m	0.711	0.711	0.756	0.913
sim_p	0.329	0.713	0.714	0.650
sim_f	0.822	0.822	0.714	0.833
SIM	0.503	0.611	0.492	**0.713**

are one feed and two products. Applying the inductive retrieval, the set composed of four source cases has been determined in the case library. Next, the global similarity is calculated for the target case and for all the source cases, using the nearest-neighborhood method. As a result, the product compositions of the target case is (0.95, 0) at the top, and (0, 0.95) at the bottom. When required, a zero element is added to the composition vector.

According to the nearest-neighborhood method (see Table 8.5), the most-similar case is a benzene–toluene problem (from Yeomans and Grossmann, 2000). However, to choose the most suitable superstructure and MINLP model, three most-similar cases are considered (cases 4, 2, and 1 in Table 8.5). In the given example, source case 1 and source case 2 have the same MINLP model (Viswanathan and Grossmann, 1993). They differ in regard to the initial point during optimization. Therefore, the adaptation of source case 3 is not studied here. The models must be adapted according to the actual requirements of the target case. The adaptation has two main steps: (1) adaptation of the model and (2) adaptation of the solutions of the source cases as an initial point.

The adaptation of the MINLP model is based on the assumptions of the optimization procedure. The column pressure assumed to be constant; therefore, the equations of the pressure profile in the model of Viswanathan and Grossmann are omitted. A constant molar overflow is assumed; therefore, the enthalpy balances and enthalpy calculations are omitted, and other equations are used instead, which force the total vapour and liquid flows to be constant in each column section.

As the heptane–toluene mixture has lower relative volatility than the mixtures of the source cases, the maximum number of trays in the column is increased to 80.

According to our earlier experiences the numerical characteristics if this kind of models can be improved by adding monotonity constraints to the model. Therefore, concentration and temperature monotonity constraints are given to the MINLP models, which do not spoil the generality of the models.

The solution of a source case is used to give the initial state in design and optimization. The number of trays in the solution of the most-similar source case is 55, the reflux ratio is 1.77, and the column diameter is 0.56 m. As in the target case, the feed is different from the feed of the source case (100 kmol h^{-1} instead of 150 kmol h^{-1}), so the values of these quantities must be modified in the initial state. Because of the lower relative volatility of the mixture in the target case, the reflux ratio and the column diameter are increased (3.54 and 1.12 m, respectively), using the same number of trays (55) in the initial state. An initial column profile is calculated by dividing the mole fraction interval between the compositions of distillate and bottom product into the same number of intervals as the number of trays. The initial temperature profile of the column is calculated similarly. The initial values of all other variables are calculated from these initial values using the model equations (Farkas et al., 2006).

The solution of the second-most-similar source case12 contains the following values: the number of trays is 25, the reflux ratio is 9.01, the flow rate of the distillate is 15 kmol h^{-1}, and the flow rate of the bottom product is 85 kmol h^{-1}. In this solution, a low number of trays is used with very high reflux ratio; therefore, in the initial state of the new problem, the number of trays is doubled (50), and the reflux ratio is diminished to 4.50, as in the solution of the source case. The purity requirements for the distillate and for the bottom product for the main component are the same, therefore, the initial value of the distillate and the bottom product are the same: 50 kmol h^{-1}. An initial column profile for concentration and temperature is calculated using the same method as in the first case.

8.7 Case Representation of Forced Unsteady State Reactor Model

The forced unsteady state reactor and processes are highly complex. Structuring the features in this case requires increased attention in order to be avoided the loss of essential characteristics and information that could influence the retrieval of similarity and the appropriate solution suggestion.

In order to obtain relevant information in the case library have been stored significant literature cases dealing with processes involving forced unsteady state operation.

The input information in case of unsteady state reactor operation analysis consists of the numerical value representation of the technical characteristics and implicit information, and some textual aspects related to the description of the problem.

The current problem is defined by the list of the features with their values. Expert opinion was used in order to set up the most important feature for the correct identification of the forced unsteady state systems statement. The data of each case representation are schematized as follows:

- Reactants: chemical class, substances names;
- Process type: combustion, oxidation, hydrogenation, reduction, synthesis, etc.;
- Reactor characteristics: type, shape, reactor and inert zone dimensions,
- Process and operating parameters: feed temperature, catalyst initial temperature, switching time, space velocity, pollutant concentration;
- Catalyst description: catalyst type, shape, particle size;
- Model description: type and dimension of mathematical model, type of transport phenomena considered, numerical method used, supplementary comments (information);
- Simplifications in mass and energy balance and related to the reactor type.

Last two groups of parameters build a solution part. The other part of parameter represents the problem part. However, in some situation, the simplification description is used as the problem identifier to determine more precise the model description.

Each case in the library of cases is represented by features grouped in specific classes that correspond to the forced unsteady state operation process. The list of features of the problem entity is put in Table 8.6. The model description entity is presented in Table 8.7.

The simplification description is composed of three entities related to mass balance, energy balance, and reactor type. An example of representation of each entity, which clearly represents the structure of it, is given in Fig. 8.7.

Table 8.6. Problem description of the forced unsteady-state reactor case

Feature	Type
Reaction type	Set
Pollutant name	Hierarchy
Pollutant concentration	Logical (lean/reach)
Reactor shape	Logical (tubular/spherical)
Catalyst support	Set
Reactor length (dimension)	Numeric
Catalyst type	Sequence
Length of the inert zone	Numeric
Feed temperature	Numeric
Catalyst temperature	Numeric
Switching time	Numeric
Space velocity	Numeric

Table 8.7. Description of a model of forced unsteady-state reactor (solution)

Feature	Type
Model type	Entity: qualitative features
Transport phenomena	Entity: qualitative features
Improved bed conductivity	Logical (yes/no)
Heat losses compensation	Logical (yes/no)
Good agreement with analytical results	Logical (yes/no)
Good estimation of the maximum asymptotic temperature	Logical (yes/no)
Necessity of supplementary analysis of the system	Logical (yes/no)
Complex dynamic behavior	Logical (yes/no)
Catalyst deactivation caused by temperature	Logical (yes/no)
Catalyst deactivation by other causes	Logical (yes/no)
Catalyst deactivation by water	Logical (yes/no)
Normal feeding position	Logical (yes/no)
Side feeding position	Logical (yes/no)

Mass Balance

Kind	Presence
ideal gases	yes
uniform inlet mixing	yes
pressure loss	no
homogeneous reaction	no
heterogeneous reaction	yes
mass acumulation	no
radial dispersion	no
axial dispersion	no

Energy Balance

Kind	Presence
gradients between phases	no
bulk temperature	yes
phases temperature	no
radial transport	yes
axial transport	no

Reactor Type

Kind	Presence
adiabatic	no
izothermal	no
non-izotermal	yes

Fig. 8.7. Example of simplifications description in case representation

8.8 Similarity Determination Between Unsteady-State Processes

The similarity measurement is used, as presented previously in Chap. 4, in order to retrieve analogous classes and features that satisfy the specific aspect of the target case. Only the classes and features with the same specific functional group (i.e. structure) are considered for retrieval.

The similarity measurement deals with symbolic and integer values. The determination of numerical distance is applied when comparing values of variables such as temperature, velocity, diameter, concentration, etc. The symbolic measure works with sets of features describing attributes in the case functions, i.e. it searches for similar features between two different data sets. The hierarchical measure finds the distance between tree nodes because classes contain trees composed of merged classes and/or features. Thus, the hierarchical measure determines the distance between two classes and/or features according to their tree representation.

The computation of similarity is performed by a measure of similarity obtained from the abstract description of the target case, by means of symbolic and numerical similarity measures of its features.

8.9 The Computer Tool for Model Selection of Forced Unsteady-State Reactor

The computer design supporting tool has been created as stand-alone application. It implements case-based design supporting approach and independent from Cabareen software. The tool has convenient user interface and support flexible case representation – allowing the user to change structure of problem and solution parts. That function is quite important when it is necessary to take into account simplification description in the retrieval phase to refine previously made selection of model.

The tool processes a problem description of a new forced unsteady state process matching it in comparison with all the cases in the case base. The most similar cases are retrieved and the best one is selected in two steps, the first one – tool based corresponding to the usual CBR retrieval and the second one – expert based taking into consideration the specific supplementary information or comments related to specific features in order to improve the reliability of a possible new solution.

The features values are introduced for the target process using the input form, as represented in Fig. 8.8 (lower image), and used for the retrieval of similar cases from the case library. The result is ranked according to degree of similarity of retrieved cases to the target case (Fig. 8.8 – main window).

Due to the complexity of forced unsteady state processes the inclusion of adaptation and evaluation of the application stages within the CBR tool would imply the integration of a complex knowledge-based system and a chemical

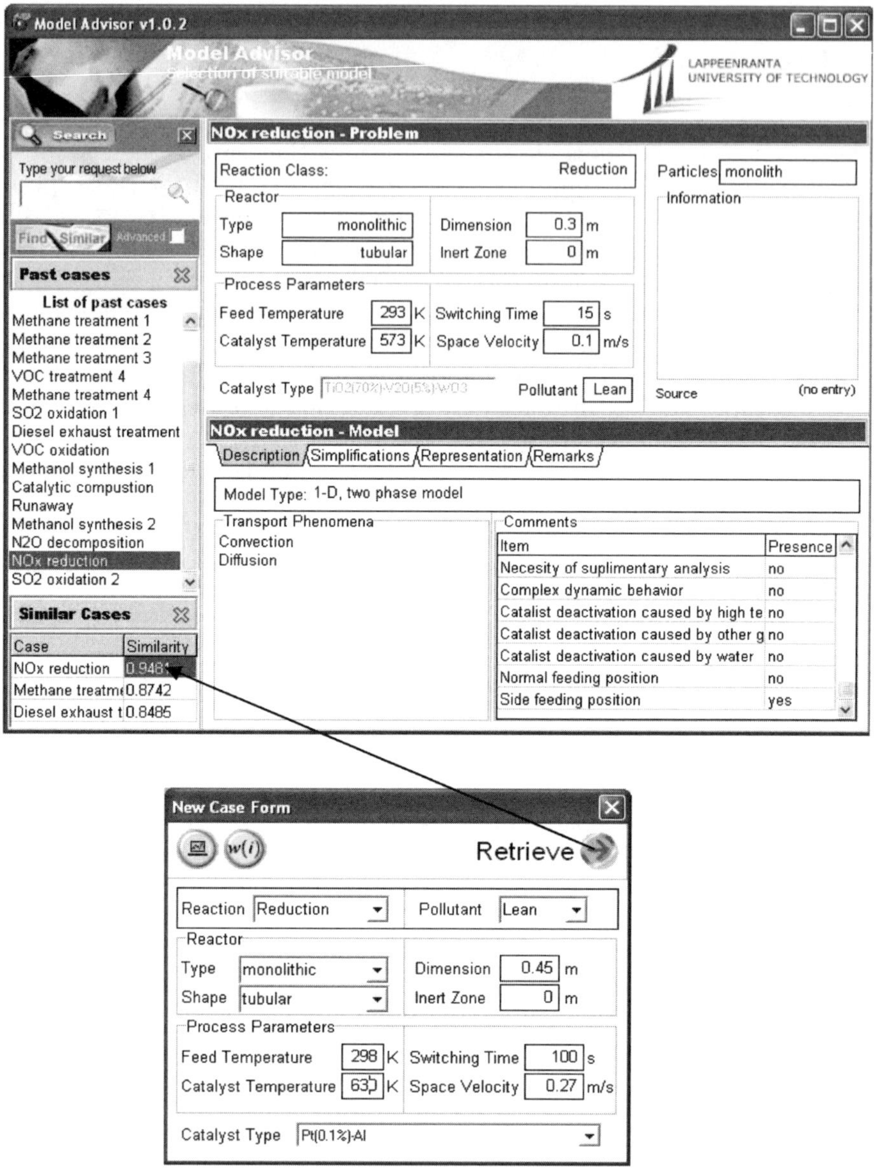

Fig. 8.8. The interface of Case-based design tool for model selection

process simulator. The objective of the present CBR tool is to support the user in the generation of process alternatives and not to carry out such generation autonomously. Even so, the evaluation of the suggested solutions is realized by using of external simulation package.

8.10 Example: Selection of Model for Catalytic Reduction of NOx with Ammonia

The objective was to select the model to the process of selective catalytic reduction of NOx with ammonia in forced unsteady-state reactor. The process parameters were represented by the following features organized as in the Table 8.8.

The maximum number of accepted case to be retrieved was fixed at the three most similar ones. The CBR tool retrieved three cases based on the information given and on the level of acceptance imposed (three cases). The problem statement in the retrieved cases is exemplified in the Table 8.9.

The most similar case found in the case library presented a degree of similarity of 0.9481 and was identified in the case library under the name "NOx reduction", case 1 in the Table 8.9. The CBR tool was designed to display all data characterizing the retrieved cases. These supplementary data is related to model description, transport phenomena involved and simplifications in mathematical model.

The use of the CBR tool provided information about the mathematical model description, degree of assumption and phenomena that contribute to the overall process behaviour in the retrieved case. The possible solutions obtained using the CBR tool have been taken into consideration and the one accepted, after expert opinion, was that provided by the most similar

Table 8.8. Problem description of the exemplary case

Feature	Value
Reaction type	Reduction
Pollutant name	Nitrogen oxides (NOx)
Pollutant concentration	Lean (order of ppm)
Reactor shape	Tubular
Catalyst support	Monolithic
Reactor length (dimension)	0.45 m
Catalyst type	Pt/Al
Length of the inert zone	0 m
Feed temperature	298 K
Catalyst temperature	630 K
Switching time	100 s
Space velocity	$0.27\,\mathrm{m\,s^{-1}}$

Table 8.9. Problem descriptions of the retrieved cases

Feature	Case 1	Case 2	Case 3
Target case similarity degree	0.9481	0.8742	0.8485
Parameters			
Reaction type	Reduction	Combustion	Decomposition
Pollutant name	Nitrogen oxides	Methane	Diesel exhaust
Pollutant concentration	Lean	Lean	Lean
Reactor shape	Tubular	Tubular	Tubular
Catalyst support	Monolithic	Monolithic	Monolithic
Reactor length (dimension)	0.3 m	0.5 m	0.58 m
Catalyst type	$TiO_2/V2O_5/WO_3$	Al2O3/Pd	$TiO_2/V2O_5/WO_3$
Length of the inert zone	0 m	0 m	0 m
Feed temperature	293 K	298 K	298 K
Catalyst temperature	573 K	400 K	298 K
Switching time	15 s	1 s	50 s
Space velocity	$0.1\,m\,s^{-1}$	$0.1\,m\,s^{-1}$	$0.1\,m\,s^{-1}$
Model type	1-D, two phase model	1-D, two phase model	3-D, two phase model
Transport phenomena	Convection Diffusion	Convection Diffusion	Convection Diffusion
Comments			
Improved bed conductivity	No	No	No
Heat losses compensation	No	No	No
Good agreement with analytical results	Yes	Yes	Yes
Good estimation of the max. asymptotic temperature	Yes	Yes	Yes
Necessity of supplementary analysis of the system	Not specified	Not specified	Not specified
Complex dynamic behaviour	Not specified	Not specified	Not specified

(continued)

Table 8.9. (*Continued*)

Feature	Case 1	Case 2	Case 3
Target case similarity degree	0.9481	0.8742	0.8485
Catalyst deactivation caused by temperature	No	No	No
Catalyst deactivation by other causes	Not specified	Not specified	Not specified
Catalyst deactivation by water	Not specified	Not specified	Not specified
Normal feeding position	No	Yes	Yes
Side feeding position	Yes	No	No
Simplifications			
Ideal gases	Yes	Yes	Yes
Uniform inlet mixing	Yes	Yes	Yes
Heterogeneous reaction	Yes	No	Yes
Bulk temperature	Yes	Yes	Yes
Isothermal system	Yes	No	Yes

retrieved case. This solution is presented as entities of Model type, Transport phenomena, and Simplification of the case 1 in the Table 8.9.

All this information is obtained in a simplified manner provided by CBR tool enabling the expert to forward reasoning about the way of dealing with a specified problem. In the present analysis of forced unsteady-state reactor operation in the case of selective catalytic reduction of NOx with ammonia, the solution provided by the information retrieved suggests that:

- The catalyst used could be the one containing $TiO_2/V2O_5/WO_3$, disposed on monolithic supports.
- The inert catalyst section used for the recuperation of the heat release during reaction could be absent.
- The feeding of the gas at normal ambient temperature does not affect the process.
- The range of initially catalyst temperature could be comprised between 400 and 600 K depending of the catalyst used.
- The reaction could be considered heterogeneous.
- The process could be described by a 1-D two phase model without affecting the reliability of the results.

Nevertheless, the final decisions are not taken at the end of the retrieved process. Just the possible solutions are suggested in this way but their reliability must be tested in the adaptation and evaluation of the application stage.

Neither adaptation nor verification can be performed automatically on the CBR tool because the suggested solution corresponds to real items and real processes. The modifications made by users to some sections during tool exploitation may affect the global performance of the process. Adaptation is highly domain dependent and it requires verification of the solution performance. Only rigorous numerical simulation can predict such performance with an acceptable accuracy. The adaptation and verification are the steps from an iterative and interactive cycle where the human designer checks the performance of the proposed cases. The iterative process finishes when the alternative solution satisfies the new requirements.

9

Equipment Design: Reactive Distillation Column Design

9.1 Introduction

The design of reactive distillation systems is considerably more complex than that of conventional reactors and distillation columns. It includes several steps (Malone and Doherty, 2000): feasibility analysis, conceptual design, equipment selection and design, operability and control studies. These steps can be corresponded to the stages of the process design model presented in this work – ABstract design (feasibility analysis), Conceptual design, DEtailed design (equipment selection and design) and Final design (operability and control studies). The methodology of case-based design support has been applied to the DE phase of the design of a reactive distillation column.

The development of column internals for a new reactive distillation application is usually based on complicated modelling and carrying out of expensive and time-consuming sequences of laboratory and pilot plant experiments. To avoid this and speed-up the design process, the computer assistant supporting the equipment design in reactive distillation is proposed.

The objective is to provide data on detailed features and geometric properties of column packing for a given process specifications and reaction description. Based on characteristics of process and catalysts the details specification of packing is selected (Fig. 9.1).

9.2 Representation of Design Case

A column design case has been presented as a set of attribute-value pairs. Each case is described as a set of features and each feature has a value. The features define the structure of the case description; the values identify the information specific to one case.

The set of the essential parameters for the correct identification of the column internals design has been selected based on the opinions of experts.

Y. Avramenko and A. Kraslawski: *Case-Based Design*, Studies in Computational Intelligence (SCI) **87**, 153–163 (2008)
www.springerlink.com

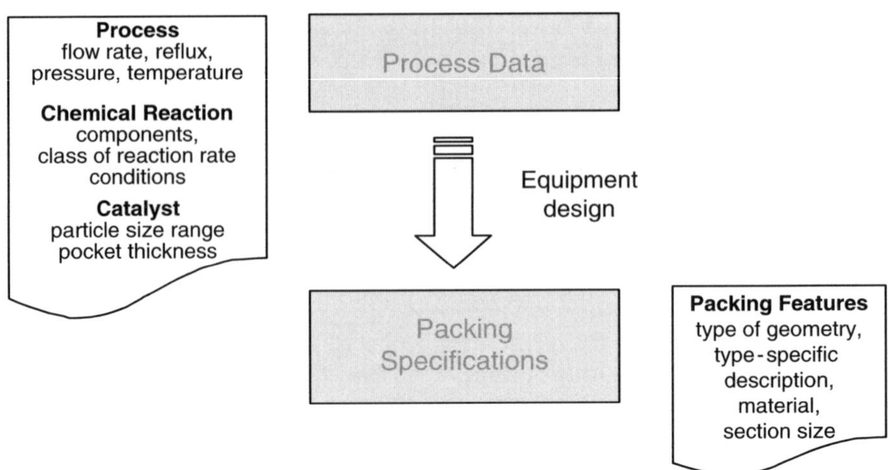

Fig. 9.1. Problem statement of selection of column internal for reactive distillation

Table 9.1. Process parameters describing problem

Reaction Description	Process and Operation	Catalyst Properties
Set of reactants	Feed flow rate, kg h^{-1}	Type of catalysis
Set of products	Product flow rate, kg h^{-1}	Granules size range, mm
Reaction temperature, °C	Feed composition	Pocket thickness, mm
Reaction pressure, bar	Product composition	Total mass, kg
Class of reaction rate	Reflux ratio	Code of composition
		Porosity
		Support (material)

Table 9.2. Packing features representing solution (divided according to type)

Monolithic	Corrugated Sheet	Element Properties
Shape of cells	Corrugation height, mm	Height, mm
Cell size, mm	Corrugation angle, mm	Diameter, mm
Rib size, mm	Corrugation length, mm	Other size, mm
Wall thickness, mm	Mesh size, mm	Catalyst vol. fraction, %
	Wire thickness, mm	Surface area, m^2 m^{-3}
	Sandwich thickness, mm	Material

Next, it has been divided into four parameters: reaction description, process parameters, catalyst description, and packing features.

The experience from the design of the internals for reactive distillation is stored as cases. The case is composed of the problem part including the description of the actual process and chemical system (Table 9.1), and the solution part containing packing features (Table 9.2).

The values of the problems part are used to identify the appropriate design combination from past experience. Feasibility and process design issues must be considered before starting to select suitable column internal. The structure of the solution part depends on the type of internal but has a common part (element properties) describing the internal element as a whole object.

Most features describing a case are of single numeric type (see previous chapter). Reaction rate is classified into qualitative values slow, medium, and fast. Code of composition has unique value for each composition and therefore can be considered a set with one element (see previous chapter). The set of reactants and set of products are the features of composite values (see previous chapter). Each substance is represented by a dangling node in the similarity tree of chemical compounds. The root of the tree represents all substances. The first-level nodes in the tree correspond to a class of chemical compounds (Organic/Inorganic). The daughter nodes correspond to subclasses of chemical substances (hydrocarbons, acids, etc.).

The design experience stored in the cases is composed of process data from US patents and commercial packing of different structures (monolithic, sandwich-like bed, modular). The case base includes cases of the production of methyl acetate, butyl acetate, methyl tertiary butyl ether (MTBE), and tert-amyl methyl ether (TAME).

9.3 Description of Decision Supporting System

A decision supporting system has been designed for pre-selection of column internals in reactive separation processes. The system helps an engineer to make a proper choice of internals type and roughly estimate the geometry, using existing experience in design of reactive separation processes. It can be used in the start phase of reactive distillation column design to determine preliminary packing specifications. The proposed geometry of the column internals can then be validated by means of a computational fluid dynamics tool for the simulation of fluid flow through the packing section.

The main tasks of the system are the following:

- To support the design of internals for a reactive distillation column by delivering design recommendations concerning the packing type;
- To store design data of internal and process descriptions of a reactive distillation column.

The structure of the system is shown in Fig. 9.2. The system consists of the following components: decision-supporting module (DSM), applied case-based reasoning to get the column packing recommendation for the new design problem, Similarity Measurements Editor (SME) and Case Base Editor (CBE) for maintaining the historical design data.

The DSM implements the reasoning procedure for the design support. A user can introduce a new problem description into the system, edit attribute specifications and corresponding weights of importance, and get a

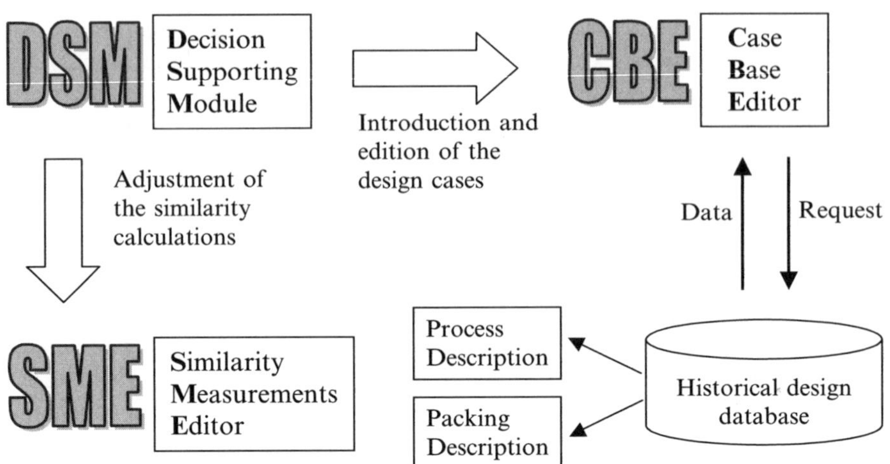

Fig. 9.2. The structure of decision supporting system for pre-selection of column internals for reactive separation

recommendation for column packing. There is a possibility to run other modules from the DSM navigation panel The reasoning method can be precisely adjusted in the SME by selecting the appropriate way of similarity calculation for past design cases. Using this module, the user is able to set a similarity value between different groups of chemical compounds through a Similarity Tree Editor. A CBE handles the various design data, stores and edits the old design cases, and introduces new design situations. The process descriptions and packing specifications are stored separately and linked in the case description. The module is also responsible for learning if the proposed packing type and specification has been proved.

The case base is organized into manageable structure that supports efficient search and retrieval methods. The cases are stored within conventional relational database structure. All data about cases have been divided on four several tables according to thematic group of parameters (such as chemical description, process parameter etc.).

Information stored within a case is of two types: indexed information that is used for retrieval; and unindexed information that may contain any type of data and is not used directly in retrieval, this information serves for additional description of the case for the user.

It is possible to work simultaneously on several problems. All data regarding problem description and the solution found are independent and stored separately.

9.3.1 Decision Supporting Module

Decision Supporting Module is able to perform the following operations: introduction of a new problem, setting of the weights of importance for each feature (attribute), getting of the recommendation about packing features, creation of the report of recommended packing specifications.

A new problem can be introduced into the system by filling corresponding fields in the special form (Fig. 9.3).

On the left side of the form, there is a full list of the attributes including their names, values and corresponding weights of importance. By marking of the respective boxes, select a set of attributes, which will be taken into account during matching procedure. The detailed information about attribute can be seen at the bottom of the form. It is possible to change the weights of importance of the attributes by double click on the caption of the column "Weight". Repeating of this action will save the changes of the values of the weights. The value of the weights can have only an integer number belonging to interval from 0 to 10.

A right side of the form contains the extended information about the attributes and it allows editing of their values. Pressing the buttons at top (Chemical Reaction, Process, Packing) it is possible to change a page of problem description. Last page is not available from start-up and serves to display output information regarding column internals.

After entering all necessary properties and parameters, the reasoner can be started. There are two options for the retrieval of the similar cases: to build the set of arbitrary number of the most similar cases ranked according to their similarity value, or to select only cases that have the similarity values greater than the fixed threshold. The result of the search is shown automatically in the last page of the form of an active problem. The comparison of retrieved cases could be performed by pressing arrows in the Packing page of the window, and the values of features are replaced for those from next case of retrieved set.

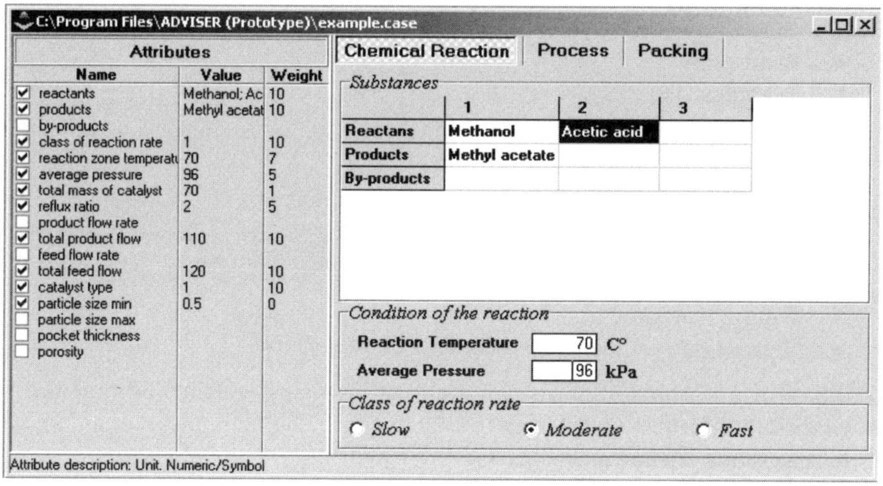

Fig. 9.3. The form for the Introduction of a new problem

All similar cases can be put into special table that is compatible with Excel format file. Thus, all retrieved data can be sent to Excel file for further validation. There is a possibility to create the report of retrieval result.

9.3.2 Case Base Editor

The case base is organized into structure supporting efficient retrieval method. The cases are stored within conventional relational database structure. All information about design cases has been divided into several parts according to thematic group of parameters: chemical reaction description (reactants, products, reaction rate etc.), process parameters (e.g. feed flow rate, reflux ration, mass of catalyst), catalyst properties (particle size, porosity, etc.), and packing features (type of internal, geometric properties). The design data is composed of process data from US patents and real industrial type of packing with different structures (monolithic, sandwich-like bed, modular).

The cases are described by several sets of parameters:

- *Chemical reaction description*
  - Reactants
  - Products
  - By-products
  - Conditions of the reaction (temperature, pressure)
  - Class of reaction rate (slow/moderate/fast)
- *Process and operating parameters*
  - Product flow
  - Feed flow
  - Reflux ratio
  - Catalyst type
- *Catalyst description*
  - Total mass
  - Granules size
  - Pocket thickness
  - Porosity
- *Detailed packing characteristics*
  - Type (structure of organization)
  - Material
  - Specific surface area
  - Volume fraction
  - Geometric specifications (depend on structure)

The editor of the design case base allows browsing, editing, adding and removing of all information relating to design case. The main panel of a manager of data sections is shown in Fig. 9.4.

A manager of data pages on left side of the screen appears when database is open. By using the manager, there is possible to show or hide all database pages included into the base.

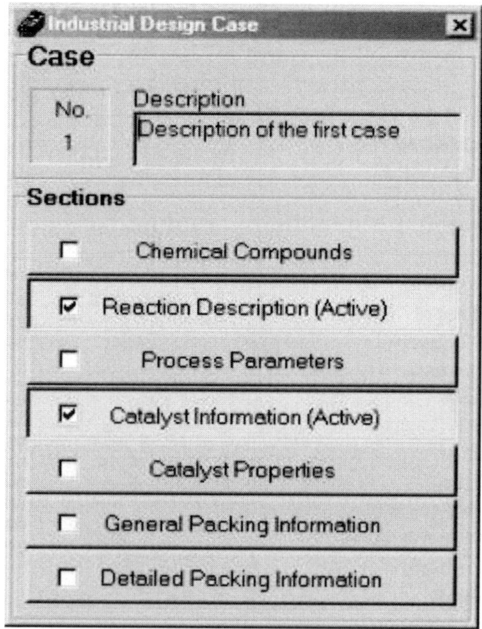

Fig. 9.4. Manager of data section of Case Base Editor

All needed information about process and packing can be introduced into the system by selecting the different data pages. The structure of the detailed packing information page varies with type of packing applied in the process. The data can be edited only if the edit mode is activated in the control panel.

The form linked with the selected data pages appears after an activation of the data page. By moving mouse cursor above the button of active data page in the manager, the corresponding data page window appears.

There are two separated data managers: for process parameters and catalyst description and for the specifications of column internals.

9.3.3 Similarity Measurement Editor

The module realizes the function of precise adjustment of similarity measures used in the retrieval procedure of DSM. It has the convenient mechanism for setting of type of local similarity functions. These functions could be adjusted using a corresponding editor. For example, the parameters, structure of similarity tree for chemical substances, the registration of the new ones as well as the definition of degree of similarity of each level can be carried out in the Similarity Tree Editor (Fig. 9.5).

Each type of data has corresponding similarity function.

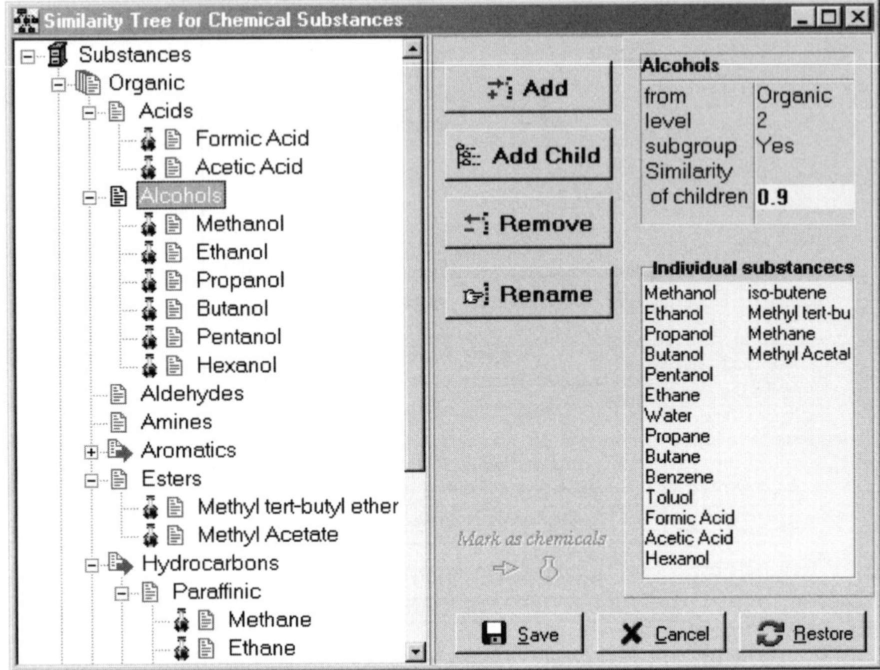

Fig. 9.5. A window of Similarity Measurements Editor

9.4 Similarity Determination

The types of data have been divided into following classes: numeric, set and hierarchical.

The difference measurements for hierarchical values have been applied to evaluate a similarity between chemical compounds basing on their chemical structure. According to this principle, so-called similarity tree, composed of the branches and nodes, was created (Fig. 9.6). The root of the tree represents all substances. The first-level nodes in the tree correspond to a basic group of the chemical compounds (Organic/Inorganic). The daughter nodes correspond to classes/subclasses of the chemical substances (hydrocarbons, aromatics, etc.).

The value of similarity between two compounds depends on the first common level where they have met. For example, methane and propane have the nearest common level "Paraffinic", but benzene and methane have the nearest common level "Organic", which means that the similarity is bigger between methane and propane than between benzene and methane. Each node in the tree has a value that allows to determine the local similarity in a numeric form, e.g. the level "organic" has a similarity 0,1, and the last level corresponding

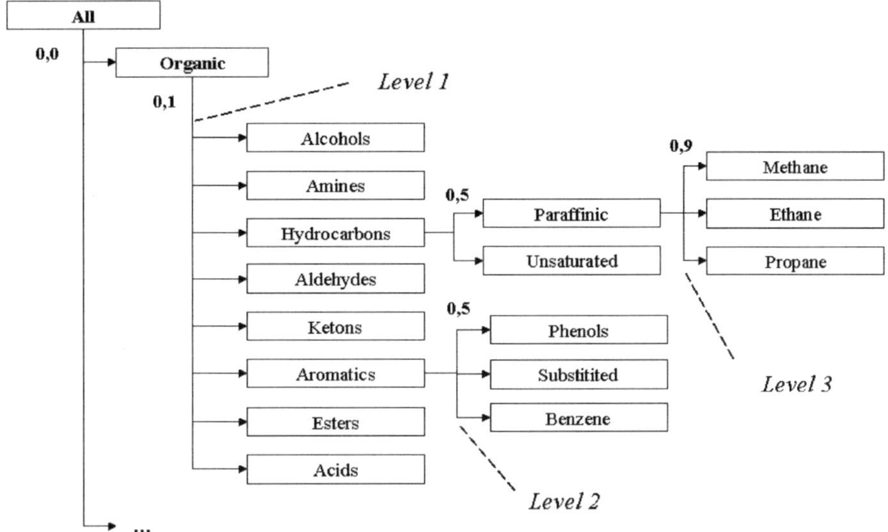

Fig. 9.6. A fragment of the similarity tree for chemical compounds

a group of most similar individual substances has the value of local similarity equal to 0,9.

Since one attribute can contain several individual compounds, the similarity of whole attribute has to be defined. The component names in the attribute can be placed in any order. Hence, in order to find two identical components in the different attribute sets there is a need to look over all elements in the sets. More general task is to find the most similar pairs of components belonging to the different sets.

For two sets A $= \{a_1, a_2, \ldots, a_n\}$ and B $= \{b_1, b_2, \ldots, b_k\}$, there is a need to find such matching $m = ((a_1, b_i), \ldots, (a_n, b_j))$ that $\sum_{(a_i,b_j)\in m} sim(a_i, b_j)$ is maximum. When the matching is found the elements of one set are rearranged to have the equal order with the most similar element from another set.

The order of the most similar pair of components is kept to be used in vector similarity measurement of composition value.

9.5 Example: Methylpropylacetate Production

An example of 2-methylpropylacetate synthesis has been selected to examine the system's applicability. The RD process for production of this industrial solvent has been introduced relatively recently. Therefore, the case library does not include industrial application of this process. At the same time there is information about suitable packing type obtained from the experiment and modelling and thus there is a possibility to evaluate the recommendation proposed by the system.

The testing task was set as follows: to select an appropriate type of packing for synthesis of 2-methylpropylacetate from 2-methylpropanol and acetic acid. To create a problem description we used the data of pilot plant experiment (Table 9.3). The class of the reaction rate was defined as moderate and general process parameters were introduced into CBR system. The reactants were introduced in the similarity tree and the similarity values between them and rest compound were determined.

The local similarity values for the numeric attributes (e.g. flow rate, temperature, etc.) were determined by the DSS. Such parameters as class of reaction rate and type of catalyst are recognized as logical type of attributes with a similarity value 1 (exact match) or 0 (not exact match). The values of weights are established basing on the experience of the designers. Very often it is treated as proprietary information of the company. In the presented example, we identified the first group of parameters (chemical reaction description) as the most important and set weights of importance to 9–10. The importance

Table 9.3. Problem description of the test case

Parameter	Value
Reactants	2-Methylpropanol; acetic acid
Products	2-Methylpropylacetate
Reaction temperature, °C	118
Reaction pressure, bar	1
Class of reaction rate	Moderate
Feed, acetic acid, kg h^{-1}	0.797
Feed, 2-methylpropanol, kg h^{-1}	1.203
Bottom product, kg h	1.5
Distillate, organic phase, kg h^{-1}	0.276
Distillate, water phase, kg/h^{-1}	0.224
Type of catalyst	Autocatalisis

Table 9.4. Detailed features of selected packing (corrugation sheet type)

Parameter	Value
Corrugation height, mm	14.9
Corrugation angle, mm	42.5
Corrugation length, mm	37
Mesh size, mm	0.5
Wire thickness, mm	0.25
Element diameter, mm	220
Element height, mm	290
Catalyst volume fraction, %	25
Surface area, m^2 m^{-3}	85

of the operating parameters has been lower (5–7), and importance of catalyst information was set to 1–3.

As a result, the system retrieved the most similar case (with the highest value of similarity) and provided the detailed information about the suitable packing type for this process. There was selected the corrugated sheet type of packing KATAPAK® manufactured by Sulzer Chemtech Ltd. (Table 9.4), the same packing that was selected as the best one during the experimental tests. Therefore, there is no need for adaptation of the proposed solution. The existing type of packing is suitable for the process under consideration.

Summary

The creation of an initial design proposal requires significant engineering experience, intuition and creativity. In order to facilitate the design process and to reduce the required time, a case-based design supporting methodology is presented in this work.

Case-based design is an approach based on the reuse of past experiences to find solutions to new, similar problems. The approach is beneficial when the problems are not completely understood and a reliable model cannot be built. It is a typical situation for most design tasks.

The results of the work can be divided into three parts:

1. Creation of a model of the design process for development of a chemical product.
   The design starts with the need to produce a chemical product with a given functionality. The overall design process is represented as three PROs – Properties design, Product design and Process Design. The identification of the physical–chemical properties of the future product based on functionality representation is called the Properties Design. The Product Design is the determination of the structural properties of the product (molecular structure, composition of mixture/blend and colloidal system). Design of the manufacturing process of a chemical product is Process Design. It evolves from different level of abstractions: abstract, basic, conceptual, details, equipment and final.
   This model gives a clear representation of steps usually passing by engineers in solving design problems. The model is useful in developing a computer assistant because it distinguishes properties, product and process design and provides a description of the objectives of each type of design.
2. A new model of Case-Based Reasoning which is flexible in the support of an evolutionary changing design task.
   The model of the CBR process is composed of six steps: collect, constitute, compile, compare, correct and check.

The first step is the collection of necessary data from the environment. Under environment is understood a set of information sources that is not part of the CBR system. Depending on the design task, appropriate specific data are extracted from the sources. The next step is to constitute the case structure that is best suited to the described area of the specific design problem to be solved. Once the case representation has been obtained the collection of relevant data is processed to create the case base of the specific case structure. During the next step, a new problem to be solved is introduced according to established case representation and compared with past cases from the created case base. Once the most similar case has been retrieved, its solution is corrected in the following step. The corrected solution is the subject of checking to be validated. The checked cases can be stored in the environment to extend its scope.

The presented model implies not only runtime reasoning but also runtime case acquisition that differs from most other CBR models. It is necessary in a changing design environment where data structure is being changed during evolution from less to more abstract levels of the design process. This advantage allows universally support of each elementary design activity described in the design model.

3. The development of a general concept of similarity which copes with diverse design data representation and development of a domain independent concept of adaptation based on the general similarity concept.

   The proposed way to build a case base which can represent diverse design data is based on consideration of the information entities. A case is a set of information entities. The number of information entities in a case may be variable. The representation of an entity is based on the assumption that any design entity can be represented by a finite set of features and relations among them. The similarity between the values of features is determined based on their difference. The measures of the degree of difference for basic types of data (vector, sets, sequences, graph) and for composite data types (set of structured elements, hierarchy with assigned values) are derived from the basic definitions of the concept.

   The proposed general similarity concept is able to cope with cases that have different structure representation in the case base and contain features expressed in different formats, as numbers, vectors, sets, sequences, graphs etc.

   An adaptation procedure is based on the assumption that not only the most similar case can be used but a set of cases located nearby the current problem in the problem space can also be used. The key assumption that a similar problem has a similar solution means that solutions of similar problems are located nearby each other. The distances between new solution and solutions of the most similar problems must correspond to the distances between the current problem and similar problems under consideration. Because the design parameters may be of various types of

data representation (combination of sets, graphs, vectors etc.), the genetic algorithm has been proposed as a global optimization method.
The adaptation method is task independent and can be applied at any design stage.

The ideas of the CBR model and general similarity concept has been implemented in a number of computer tools for the support of several design tasks: fat and oil product formulations, selections of internals of reactive distillation columns, selection of a proper model for distillation system synthesis, and conceptual design of wastewater treatment systems.

Case-based reasoning (CBR) can support innovative design and redesign activity by reminding designers of previous experiences that can match with the new design situation, not necessary totally but partially. This approach is able to support almost all steps of chemical process design, except perhaps the first and last ones (Abstract and Final designs). Some limitations to the application can be found in independent domain adaptation concepts. The use of domain knowledge is necessary in most cases to correct the solution.

The design assistants that have been developed based on the case-based design methodology can reduce design time and facilitate the design process. The methodology is applicable for different stages of development of a chemical product or a process.

Further extensions of the work are seen in the derivation of new measures of difference value for composites structures, study of the applicability of the adaptation concept, automation of checking and constitution phases of the CBR model, and cases acquisition using data mining and artificial intelligence methods.

The casebased reasoning environment (CABAREEN) developed as implementation of the described method can further be extended to be more flexible and more functional. The environment is an open product and additional modules can easily be linked to it.

References

Aamodt, A., Nygard, M., 1995. *Different Roles and mutual dependencies of data, information and knowledge*, In Data and Knowledge Engineering 16, Elsevier, Holland, pp. 191–222.

Aamodt, A., Plaza, E., 1994. *Case-based reasoning: foundational issues, methodological variations, and system approaches*, Artificial Intelligence Communications, 7, pp. 39–59.

Aggarwal, A., Floudas, C.A., 1992. *Synthesis of heat integrated non-sharp distillation sequences*, Computers and Chemical Engineering, 16 (2), pp. 89–108.

Aha, D.W., 1991. *Case-Based Learning Algorithm*, in Bareiss, R. (ed.), Proceedings of. Case-based reasoning workshop, Morgan Kaufmann, pp. 147–158.

Aha, D., Breslow, L.A., Munoz-Avila, H., 1999. *Conversational case-based reasoning*, Journal of Artificial Intelligence, 2.

Allen, B.P. 1994. *Case-based reasoning: business applications*, Communications of the ACM, 37 (3), pp. 40–42.

Alterman, R., 1988. Adaptive planning, Cognitive Science 12, pp. 393–422.

Althoff, K.D., Auriol, E., Barletta, R., Manago, M., 1995. *A Review of Industrial Case-Based Reasoning Tools*, AI Intelligence.

Amen, R., Vomacka, P., 2001. *Case-Based Reasoning as a Tool for Materials Selection*, Material and Design, 22, Elsevier, pp. 353–358.

Ashley, K.D., 1990. *Modelling Legal Argument: Reasoning with Cases and Hypotheticals*, Cambridge, MA: MIT, Bradford Books.

Avramenko, Y., Nystrom, L., Kraslawski, A., 2002. *Selection of Internals for Reactive Distillation Column – Case-based Reasoning Approach*, in Grievink, J., Schijndel, J. (eds.), Proceedings of European Symposium on Computer Aided Process Engineering, 12, Elsevier, pp. 157–162.

Bain, W., 1986. Case-based reasoning: A computer model of subjective assessment, Ph.D. diss., Department of Computer Science, Yale University.

Baker, D., Bridges, D., Hunter, R., Johnson, G., Krupa, J., Murphy, J., Sorenson, K., 2002. *Guidebook to Decision-Making Methods*, WSRC-IM-2002-00002, Department of Energy, USA. http://emi-web.inel.gov/Nissmg/Guidebook_2002.pdf

Belskiy, A.A., 1979. *Theory of Graph and Theory of Combinations*, MIIT, Moscow (in Russian).

Bernardo, F.P., Saraiva, P.M., 2005. *Integrated Process and Product Design Optimization: A Cosmetic Emulsion Application*, in Puigjaner, L., Espuna, A. (eds.), Proceedings of European Symposium on Computer Aided Process Engineering, 15, Elsevier, pp. 1507–1512.

Bonzano, A., Cunningham, P., Smyth, B., 1997. *Using Introspective Learning to Improve Retrieval in Car: A Case Study in Air Traffic Control*, Proceedings of Second International Conference on Case-Based Reasoning, ICCBR-97, Providence RI, USA, pp. 291–302.

Botar-Jid, Ch., 2007. *Selective catalytic reduction of nitrogen oxides with ammonia in forced unsteady state reactors: Case based and mathematical model simulation reasoning*, PhD Thesis, Lappeenranta University of Technology, Finland.

Brachman, R., Schmolze, J., 1985. *An Overview of the KL-ONE Knowledge Representation System*, Cognitive Science, 9 (2), pp. 171–216.

Braha, D., Maimon, O., 1998. *A Mathematical Theory of Design: Foundations, Algorithms and Applications*, Kluwer, Dordrecht, The Netherlands.

Brown, D., Chandrasekaran, B., 1985. *Expert System for a class of mechanical design activity*, in Gero, J. (ed.), Knowledge Engineering in Computer-Aided Design, Amsterdam, North Holland.

Burkhard, H.D., 1998. *Extending Some Concepts of CBR – Foundations of Case Retrieval Nets*, in Lenz, M., Bartsch-Sporl, B., Burkhard, H.D., Wess, W. (eds.), Lecture Notes in Artificial Intelligence, 1400, Springer-Verlag Berlin Heidelberg, Germany.

Caballero, J.A., Grossmann, I.E., 1999. *Aggregated model for integrated distillation systems*, Industrial and Engineering Chemistry Research, 38, pp. 2330–2344.

Caballero, J.A., Grossmann, I.E., 2001. *Generalized disjunctive programming model for the optimal synthesis of thermally linked distillation columns.* Industrial and Engineering Chemistry Research, 40, pp. 2260–2274.

Chandrasekaran, B., 1990. *Design Problem Solving: A Task Analysis*, AI Magazine, 11 (4).

Chaput, A.B., 1999. *Tackle troubleshooting with a case-based expert system*, Chemical Engineering Progress, 95 (4), pp. 57–62.

Clancey, W.J., 1985. *Heurestic classification*, Artificial Intelligence, 27, pp. 289–350.

Clark, R., Chopeta, L., 2004. *Graphics for Learning: Proven Guidelines for Planning*, Designing, and Evaluating Visuals in Training Materials, Jossey-Bass/Pfeiffer.

Coello, J.M.A., Santos, R.S., 1999. *Integrating CBR and Heuristic Search for Learning and Reusing Solutions in Real-Time Task Scheduling*, Case-Based Reasoning, Research and Development, (ICCBR99) LNCS 1650, Springer-Verlag Berlin Heidelberg, pp. 89–103.

Coyne, R.D., Rosenman, M.A., Radford, A.D., Balachandran, M, Gero, J.S., 1990. *Knowledge-Based Design Systems*, Reading, Addison-Wesley, USA.

Cross, N. (ed.), 1984. *Development in Design Methodology*, Wiley, New York.

Cross, N., 2000. *Engineering Design Methods: Strategies for Product Design*, Wiley, Chichester, UK.

Cussler, E.L., Moggridge, G.D., 2001. *Chemical Product Design*, Cambridge University Press, USA.

Dasgupta, S., 1989. *The Structure of Design Processes*, in Yovits, M.C. (ed.), Advance in Computers, 28, Academic Press, New York, pp. 1–67.

Dixon, J.R., Duffey, M.R., Irani, R.K., Meunier, K.L., Orelup, M.F., 1988. *A Proposed Taxonomy of Mechanical Design Problems*, Proceedings of ASME Computers in Engineering Conference, ASME, San Francisco, USA.

Domeshek, E.A., Kolonder. J.L., 1992, *A Case-Based Design Aid for Architecture*, in Gero, J.S. (ed.), Artificial Intelligence in Design-92, AID, Kluwer, Dordrecht, Pittsburg, pp. 497–516.

Duda, R.O., Hart, P.E., Stork, D.G., 1998. *Pattern Classification and Scene Analysis: Part I Pattern Classification*, John Wiley and Sons Inc., USA.

Duran, M.A., Grossmann, I.E., 1986. *A mixed-integer non-linear programming approach for process systems synthesis*, AIChE Journal, 32 (4), pp. 592–606.

Dym, C.L., Levitt, R.E., 1991. *Knowledge-Based Systems in Engineering*, McGraw-Hill, New York, USA.

Dym C.L., Little, P., 2004. *Engineering Design: A Project-Based Introduction*, Wiley, USA.

Emery, J., 1987. *Management Information Systems, The Critical Strategic Resource*, Oxford University Press, New York, USA.

Farkas, T., Avramenko, Y., Kraslawski, A., Lelkes, Z., Nyström, L., 2006. *Selection of a Mixed-Integer Nonlinear Programming (MINLP) Model of Distillation Column Synthesis by Case-Based Reasoning*, Industrial and Engineering Chemistry Research, 45 (6), pp. 1935–1944.

Finnie, G., Sun, Z., 2003. *R^5 Model for Case-Based Reasoning*, Knowledge-Based Systems, 16, Elsevier, pp. 59–65.

Flemming, U., Zeyno, A., Coyne, R., Snyder, J., 1997. *Case-Based in Design in a Software Environment that Supports the Early Phases*, in Maher, M.L., Pu, P. (eds.), Lawrence Erlbaum Associates, Mahwah, USA, pp. 61–86.

Freitas, I.S.F., Costa, C.A.V., Boaventura, R.A.R., 2000. *Conceptual design of industrial wastewater treatment process: primary treatment*, Computers and Chemical Engineering., 24, pp. 1725–1730.

French, M.J., 1985. *Conceptual Design for Engineers*, Design Council, London.

French, M.J., 1992. *Form, Structure and Mechanism*, MacMillan, London.

Gachet, A., 2004. *Building Model-Driven Decision Support Systems with Dicodess*, Zurich, VDF.

Gani, R., 2004. *Chemical product design: challenges and opportunities*, Computers and Chemical Engineering, 28, pp. 2441–2457.

Gebhardt, F., Voss, A., Grather, W., Schmidt-Belz, B., 1997. *Reasoning with Complex Cases*, International Series in Engineering and Computer Science, 393, Kluwer, Boston.

Gero, J.S., 1990. *Desing Prototypes: A Knowledge Representation Schema for Design*, AI Magazine, 11 (4), pp. 26–36.

Goel, A., Chandrasekaran, B., 1992. Case-based design: A task analysis. In Artificial intelligence approaches to engineering design, vol. 2: Innovative design, ed. C. Tong and D. Sriram, Academic Press.

Grabowski, H., Lossack, R.-S., Weis, C., 1995. *Supporting the Design by an Integrated Knowledge-Based Design System*, in Gero, J., Sudweeks, F. (eds.), Proceedings of IFIP WG5.2 Workshop on Formal Design Methods for Computer-Aided Design, Chapman and Hall, London.

Grossmann, I.E., 1985. *Mixed-integer programming approach for the synthesis of integrated process flowsheets*, Computers and Chemical Engineering, 20, pp. 655–662.

Haag, S., Cummings, M., McCubbrey, D., Pinsonneault, A., Donovan, R., 2006. *Management Information Systems for the Information Age* (3rd Canadian Ed.), Canada, McGraw Hill Ryerson.

Hackathorn, R.D., Keen, P.G.W., 1981. *Organizational Strategies for Personal Computing in Decision Support Systems.* MIS Quarterly, 5 (3).

Harris, R., 1998. Introduction to Decision Making, VirtualSalt. *http://www.virtualsalt.com/crebook5.htm*

Häettenschwiler, P., 1999. *Neues anwenderfreundliches Konzept der Entscheidungsunterstützung.* Gutes Entscheiden in Wirtschaft, Politik und Gesellschaft. Zurich, vdf Hochschulverlag AG: 189–208.

Heckerman, D., 1991. *Probabilistic Similarity Networks*, MIT, Cambridge.

Heider, R., Auriol, E., Tartarin, E., Manago, M., 1997. *Improving the Quality of Case Bases for Building Better Decision Support Systems*, in Bergmann, R., Wilke, W. (eds.), 5th German Workshop on CBR – Foundations, Systems, and Applications, Report LSA-97-01, Kaiserslautern, University of Kaiserslautern, pp. 85–100.

Hennessy, D., Hinkle, D., 1991. *Initial Results from Clavier: A Case-Based Autoclave Loading Assistant*, in Bareiss, R. (ed.), Proceedings of Case-based Reasoning Workshop, Morgan Kaufmann, pp. 225–232.

Holsapple, C.W., Whinston, A.B., 1996. *Decision Support Systems: A Knowledge-Based Approach*, St. Paul, West Publishing.

Hunt, J., 1995. *Evolutionary Case Based Design*, in Waston, I.D. (ed.), Progress in Case-Based Reasoning, LNAI 1020, Springer-Verlag Berlin Heidelberg, pp. 17–31.

King, J.M.P., Banares-Alcantara, R., Manan, Z.A., 1999. *Minimising environmental impact using CBR: an azeotropic distillation case study*, Environmental Modelling and Software, 14 (5), pp. 359–366.

Kolonder, J.L., 1984. *Retrieval and Organization Strategies in Conceptual Memory*, Lawrence Erlbaum, Hillsdale, USA.

Kolonder, J.L., 1993. *Case-Based Reasoning.* Morgan Kaufmann, San Mateo, USA.

Koton, P., 1989. *Evaluating Case-Based Problem Solving*, in Kolonder, J.L. (ed.), Proceedings of Case-Based Reasoning Workshop, Morgan Kaufmann, pp. 173–175.

Kovacic, K., Sterling, L., Petot, G., Ernst, G., Yang, N., 1992. *Towards an Intelligent Nutrition Manager*, Proceedings of ACM/SIGAPP Symposium on Computer Applications, ACM, pp. 1293–1296.

Kraslawski, A., Lyssov, I., Kudra, T., Borowiak, M., Nystrom, L., 1999a. *Case-based reasoning for equipment selection using rough sets analysis in adaptation phase*, Computers and Chemical Engineering, 23 (Suppl.), Elsevier, pp. 707–710.

Kraslawski, A, Pedrycz, W., Nystrom, L., 1999b. *Fuzzy Neural Network as Instance Generator for Case-based Reasoning System: An Example of Selection of Heat Exchange Equipment in Mixing Tanks*, Natural Computing and Application, 8, Springer Berlin Heidelberg, London, pp.106–113.

Krovvidy, S., Wee, W., Suidan, M., Summers, R., Coleman, J., 1994. *Intelligent sequence planning for wastewater treatment systems*, IEEE Expert, 9, pp. 15–20.

Lansdown, J., 1987. *Design studies*, 8 (2), pp. 76–81.

Lau, H.C.W., Wong, C.W.Y., Hui, I.K., Pun, K.F., 2003. *Design and implementation of an integrated knowledge system*, Knowledge-Based Systems, 16 (2), pp. 69–76.

Lenz, M., Auriol, E., Manago, M., 1998. *Diagnosis and Decision Support*, in Lenz, M., Bartsch-Sporl, B., Burkhard, H.D., Wess, S. (eds.), Lecture Notes in Artificial Intelligence, 1400, Springer Berlin Heidelberg, Germany.

Li, X., Kraslawski, A., 2004. *Conceptual Process Synthesis: Past and Current Trends*, Chemical Engineering and Processing, 43 (5), pp. 589–600.

Liew, P.S., Gero, J.S., 2002. *An Implementation Model of Constructive Memory for a Situated Design Agent*, in Gero, J.S., Brazier, F. (eds.), Agents in Design, Key Centre of Design Computing and Cognition, University of Sydney, Australia, pp. 257–276.

Maher, M.L., Pu, P., (eds.), 1997. *Issues and Applications of Case-Based Reasoning in Design*, Lawrence Erlbaum Associates, Mahwah, USA.

Malone, M.F., Doherty, M.F., 2000. *Reactive Distillation*, Industrial and Engineering Chemistry Research, 39, pp. 3953–3957.

Marakas, G.M., 1999. *Decision Support Systems in the Twenty-First Century*, Upper Saddle River, N.J., Prentice Hall.

Mejasson, P., Petridis, M., Knight, B., Soper, A., Norman, P., 2001. *Intelligent design assistant (ITA): a case base reasoning system for material and design*, Material and Design, 22, Elsevier, pp. 163–170.

Mileman, T., Knight, B., Petridis, M., Cowell, D., Ewer, J., 2002. *Case-based retrieval of 3-D shapes for the design of metal castings*, Journal of Intelligent Manufacturing, Vol. 13(1), Kluwer.

Minsky, M.A., 1981. *Framework for Representing Knowledge*, Mind Design, MIT, pp. 95–128.

Moorman, K., Ram, A., 1992. *A Case-based approach to reactive control for autonomous robots*, In Proceedings of the AAAI Fall Symposium on AI for Real-World Autonomous Robots, Cambridge, AAAI Press.

Mostow, J., Barley, M., Weinrich, T., 1989. *Automated reuse of design plans*, International Journal for Artificial Intelligence in Engineering 4(4), pp. 181–196.

Munoz-Avila, H., Weberskirch, F., 1996. *A specification of the domain of process planning: properties, problems and solution*, Technical report LSA-96-10E, Centre for Learning Systems and Applications, University of Kaiserslautern, Germany.

Nakayama, T., Tanaka, K., 1999. *Computer-assisted thermal analysis system founded on case-based reasoning*, Journal of Chemical Information and Computer Science, 39, American Chemical Society, pp. 819–832.

Narashiman, S., Sycara, K., Navin-Chandra, D., 1997. *Representation and Synthesis of Non-Monotonic Devices*, in Maher, M.L., Pu, P., (eds.), Lawrence Erlbaum Associates, Mahwah, USA, pp. 187–220.

Novak, Z., Kravanja, Z., Grossmann, I.E., 1996. *Simultaneous synthesis of distillation sequences in overall process schemes using an improved MINLP approach*, Computers and. Chemical Engineering, 20 (12), Elsevier, pp. 1425–1440.

O'Brien, R.D., 2004. *Fats and oils: formulation and processing for applications*, CRC Press, New York, USA.

Pahl, G., Beitz, W., 1984. *Engineering Design*, Design Council Books, London.

Pajula, E., Seuranen, T., Hurme, M., 2001. *Synthesis of separation processes by using case-based reasoning*, Computers and Chemical Engineering, 25, Elsevier, pp. 775–782.

Porter, B.W., Bareiss, E.R., 1986. *PROTOS: An experiment in knowledge acquisition for heuristic classification tasks*, In, Proceedings of the First International Meeting on Advances in Learning (IMAL), pp. 159–174.

Power, D.J., 1997. *What is a DSS* ? The On-Line Executive Journal for Data-Intensive Decision Support, 1 (3).

Power, D.J., 2002. *Decision support systems: concepts and resources for managers*. Westport, Conn., Quorum Books.

Purvis, L., Pu, P., 1995. *Adaptation Using Constraint Satisfaction Techniques*, in Aamodt, A., Veloso, M. (eds.), Case-Based Reasoning and Development, Proceedings ICCBR-95, Lecture Notes in Artificial Intelligence, 1010. Springer-Verlag Berlin Heidelberg, pp. 289–300.

Richter, M.M., 1992. *Prinzipien der Kunstlichen Intelligenz*, B.G. Teubner, Stuttgart, Germany.

Rinderle, J.R., 1987. *Function and Form Relationships: A Basis for Preliminary Design*, Report EDRC-24-05-87, Carnegie Mellon University Engineering Design Research Center, Pittsburgh: USA.

Rivard, H., Fenves, S.J., 2000. *SEED-Config: a case-based reasoning system for conceptual building design*, Artificial Intelligence for Engineering Design, Analysis and Manufacturing.

Roda, I.R., Poch, M., Sanchez-Marre, M., Cortes, U., Lafuente, J., 1999. *Consider a case-based system for control of complex processes*, Chemical Engineering Progress, 95 (6), pp. 39–45.

Roozenburg, N.F.M., Eekels, J., 1995. *Product Design: Fundamentals and Methods*, Wiley, Chichester, New York.

Rousu, J., Aarts, R.J., 1996. *Adaptation costs as a criterion for solution evaluation*, in Smith, I., Faltings, B. (eds.), Advances in Case-Based Reasoning, Lecture Notes in Artificial Intelligence, 1186, Springer-Verlag Berlin Heidelberg, pp. 354–361.

Sanchez-Marre, M., Cortes, U., Roda, I.R., Poch., M., Lafuente, J., 1997. *Learning and Adaptation in Wastewater Treatment Plants through Case-Based Reasoning*, Microcomputers in Civil Engineering, 12 (4), pp. 251–266.

Schalkoff, R.J., 1992. Pattern recognition: Statistical, structural, and neural approach. John Wiley and Sons Inc.

Schank, R.C., 1982. *Dynamic Memory: A Theory of Learning in Computers and People*, Cambridge University Press, New York.

Schwartz, A.B., Barcia, R.M., Martins, A., Weber-Lee, R., 1997. *PSIQ – A CBR Approach to the Mental Health Area*, in Bergmann, R., Wilke, W. (eds.), 5th German Workshop on CBR – Foundations, Systems, and Applications, Report LSA-91-01E, Kaiserslautern, University of Kaiserslautern, pp. 217–224.

Seider, W.D., Seader, J.D., Lewin, D.R., 1999. *Process Design Principles: Synthesis, Analysis and Evaluation*, Wiley, USA.

Seider, W.D., Seader, J.D., Lewin, D.R., 2004. *Product and Process Design Principles: Synthesis, Analysis, and Evaluation*, Wiley, USA.

Siddall, J.N., 1982. *Optimal Engineering Design: Principles and Applications*, New York, Dekker.

Sprague, R.H., Carlson, E.D., 1982. *Building effective decision support systems*, Englewood Cliffs, N.J., Prentice-Hall.

Stanfill, C., Waltz, D., 1986. *Toward memory-based reasoning*, Communications of the ACM, 29 (12), pp. 1213–1228.

Stanhope, P., 2002. *Get in the Groove: building tools and peer-to-peer solutions with the Groove platform*, New York, Hungry Minds.

Suh, M.S., Jhee, M.C., Ko, Y.K., Lee, A., 1998. *A case-based expert system approach for quality design*, Expert Systems with Applications, 15, Pergamon, pp. 181–190.

Sun Z., Finnie G., Weber K., 2004. Case Base Building based on Similarity Relations. *Information Science*, 165, pp. 21–43.

Surma, J., Braunschweig, B., 1996. *case-base retrieval in process engineering: supporting design by reusing flowsheets, engineering*, Application of Artificial Intelligence, 9(4), pp. 385–391.

Sycara, E.P., 1987. *Finding creative solutions in adversarial impasses*, In Proceedings of the Ninth Annual Conference of the Cognitive Science Society, Northvale, Erlbaum.

Tong, C., Sriram, D. (eds.) 1992. *Artificial Intelligence in Engineering Design*, Boston, Academic Press.

Townsend, D.W., Linnhoff, B., 1983. *Heat and Power Networks in Process Design; Part I and II*, AIChE Journal, 29, pp.742–748.

Tsatsoulis, C., Alexander, P., 1997. *Integrating Cases, Subcases, and Generic Prototypes for Design*, in Maher, M.L., Pu, P. (eds.), Lawrence Erlbaum Associates, Mahwah, USA, pp. 261–300.

Turban, E., Aronson, J.E., Liang, T.P., 2005. *Decision Support Systems and Intelligent Systems.* New Jersey, Pearson Education, Inc.

Turton, R., Bailie, R.C., Whiting, W.B., Shaewitz, J.A., 1998. *Analysis, Synthesis, and Design of Chemical Processes*, Prentice Hall PTR, USA.

Ulrich, K.T., Eppinger, S.D., 2000. *Product Design and Development*, McGraw-Hill Companies, Inc., USA.

Viswanathan, J., Grossmann, I.E., 1993a. *An alternate MINLP model for finding the number of trays required for a specified separation objective.* Computers and Chemical Engineering, 17 (9), pp. 949–955.

Viswanathan, J., Grossmann, I.E. 1993b. *Optimal feed locations and number of trays for distillation columns with multiple feeds.* Industrial and Engineering Chemistry Research, 32, pp. 2942–2949.

Voss, A., 1995. *Similarity concepts and retrieval methods*, FABEL Report No. 13, Gesellschaft fur Mathematik und Datenverarbeitung mbH, Santa Cruz.

Voss, A., 1997. *Case Design Specialists in FABEL*, in Maher, M.L., Pu, P., (eds.), Lawrence Erlbaum Associates, Mahwah, USA, pp. 301–338.

Walas, S.M., 1988. *Chemical Process Equipment*, Butterworth Publishers, USA.

Watson, I., Gardingen, D.A., 1999. *Distributed case-based reasoning application for engineering sales support*, Proc. 16th International Joint Conference on Artificial Intelligence (IJCAI-99), 1, Morgen Kaufmann, San Francisco, CA, pp. 600–605.

Wibowo, C., Ng, K.M., 2001. *Product-oriented process synthesis and development: creams and pastes*, AIChE Journal, 47 (12), pp. 2746–2767.

Wilke, W., Lenz, M., Wess, S., 1998. *Intelligent Sales Support with CBR*, in Lenz, M., et al. (eds.), Case Based Reasoning Technology, from Foundations to Applications, LNCS 1400, Springer-Verlag Berlin Heidelberg, pp. 91–113.

Yang, C.T., Kao, J.J., 1996. *An expert system for selecting and sequencing wastewater treatment processes*, Water Science & Technology, 34, pp. 347–353.

Yeomans, H., Grossmann, I.E., 1999. *Nonlinear disjunctive programming models for the synthesis of heat integrated distillation sequences*, Computers and Chemical Engineering, 23 (9), pp. 1135–1151.

Yeomans, H., Grossmann, I.E., 2000. *Disjunctive programming models for the optimal design of distillation columns and separation sequences*, Industrial and Engineering Chemistry Research, 39 (6), pp. 1637–1648.

Appendix

Table A.1. List of commands and descriptions of their functions

Command	Key	Parameter	Descriptions
'		*comment*	Put a comment line on the message screen
echo		on	All messages and command traces are displayed in the message screen
		off	All messages are denied to display
load		*filename*	Load a file with the name specified by parameter *filename* to the environment. The environment is able to work only with one loaded file.
edit		[*filename*]*	Load a file with specific *filename* to internal text editor and open editor window. If parameter is omitted then the command sends to editor the previously loaded file.
new			Create a new text file and open empty edit window
save		[*filename*]	Save to a file with *filename* the content of the editor. Without parameter the command saves the work file of the environment (previously loaded).
clear			Clear the message screen
exe		*filename*	Execute the script stored in a file with *filename*. The extension of a file can be omitted; in this case, the extension .exl (executable lines) is automatically added to a filename.
dbinit			Initialization of a database. The specifications of the database must be loaded first from the file of specific format (*data source description* file).

(continued)

Table A.1. Continued

Command	Key	Parameter	Descriptions
crtables			Create internal objects related to tables of initialized relational database. Specifications for tables are taken from data source description file.
	d	short	Display short details about creating tables
	d	extended	Display extended details about creating tables
setrelt			Set relations between tables in relational database. The information about relations is taken from data source description file.
datacnt			Connect to the initialized database. The tables must be created first.
dbclose			Disconnect from the database.
ver			Display the version of the environment
help			Display list of available command
	d		Add syntax and short description for each command in displaying list.
quit			Exit from the environment (kept for compatibility with old version)
exit			Exit from the environment
pause			Stop execution of the script and return to command line or other active window of the environment.
cont			Continue execution of stopped script.
path		[*new path*]	Display work directory of the environment. If parameter is given, which is valid directory or special symbols (such as '..' for parent directory), the work directory is changed to those specified by parameter.
dirlist		[*filemask*]	Display the list of files located in the work directory of the environment. *Filemask* defines what kind of files is displayed (for example, '*.exl' shows only executable scripts files).
crfslist		[*filemask*]	Compose list of files that are source data files for case base (if the case base is complied from a set of files of certain structure). *Filemask* defines the template for files to be included in the list. If parameter is omitted then information about file base is taken from previously loaded file;
	d	[*filemask*]	Display the names of files which were added to the list.

(continued)

Table A.1. Continued

Command	Key	Parameter	Descriptions
cbinit		[*name*]	Create the structure of the cases based on specifications taken from previously loaded file of specific format – *case structure description* file. Parameter *name* specifies name of initializing case structure.
crcb	fl		Compile initiated case base (with assigned case structure) from list of files. The list must be created first. The case base is saved into a file with filename given by name of structure.
	db		Compile initiated case base from connected data base. The case base is saved into a file with filename given by name of structure
comp			Compare the new problem specified in the previously loaded file with cases in the case base. The case base must be created first. The similar case (their numbers or names) are listed on the screen;
	bs	*number*	Retrieve only *number* of best similar cases
	hs	*value*	Retrieve the cases that have similarity higher than certain *value*.
setsim		[*filename*]	Set the similarity measurement for used data types. Information is taken from a file with *filename* of specific structure (similarity description file). If parameter is omitted then previously loaded file is used;
	r	[*filename*]	Rewrite embedded similarity measurements
	a	[*filename*]	Add similarity measurements for new type of data (new type of composite values).
setga		[*filename*]	Set the specifications for genetic algorithm procedure for a file with *filename*. If parameter is omitted then previously loaded file is used.
adapt		[*number*]	Perform the procedure adaptation for the introduced problem using the set of retrieved similar cases of cardinality specified by *number*. The result is saved into file *newsol.cml*.
scale		*set size*	Scaling the solution space based on problem space with neighbourhood of solutions in solution space specified by *set size* parameter.
%		[*utilityname*]	Run the utility specified by *utilityname*. If parameter is omitted then a list of registered utilities is displayed on the message screen.
run		*filename*	Launch the interface which is separate application with a given *filename*.

* – If parameter is in the brackets [] then it can be omitted.

Printing: Krips bv, Meppel, The Netherlands
Binding: Stürtz, Würzburg, Germany